A FIELD GUIDE TO
Poisonous Plants and Mushrooms of North America

A FIELD GUIDE TO
Poisonous Plants and Mushrooms of North America

CHARLES KINGSLEY LEVY
RICHARD B. PRIMACK

Boston University

Illustrations by Laszlo Meszoly
and Margaret Huong Primack

THE STEPHEN GREENE PRESS
Brattleboro, Vermont
Lexington, Massachusetts

We are grateful to the following photographers for their permission to include their color photographs in the color illustration section: Plates 2, 7, 10, 12, 14, 15, 20, 25 © Albert Bussewitz; Plates 5, 9, 22 © Scooter Cheatham; Plates 27, 30 © Dr. Charles Hrbek; Plates 1, 3, 4, 6, 8, 11, 13, 16, 17, 18, 19, 21, 23, 31 © Frederica Matera; Plates 28, 29, 32 © William Ormerod, Jr.; Plate 24 © Karlene Schwartz; Plate 26 © Dr. Robert Wyatt.

First Edition

Text and line illustrations copyright © 1984 by The Stephen Greene Press

All rights reserved. No part of this book may be reproduced without written permission from the publisher, except by a reviewer who may quote brief passages or reproduce illustrations in a review; nor may any part of this book be reproduced, stored in a retrieval system, or transmitted in any form or by any means electronic, mechanical, photocopying, recording, or other, without written permission from the publisher.

This book is manufactured in the United States of America. It is designed by Irving Perkins Associates and published by The Stephen Greene Press, Fessenden Road, Brattleboro, Vermont 05301.

Distributed in the United States by E.P. Dutton, Inc., New York.

Library of Congress Cataloging in Publication Data

Levy, Charles K., 1924–
   A field guide to poisonous plants and mushrooms of North America.

   Bibliography: p.
   Includes index.
   1. Poisonous plants—North America—Identification.
   2. Mushrooms, Poisonous—North America—Identification.
   3. Poisonous plants—Toxicology. 4. Mushrooms, Poisonous—Toxicology. I. Primack, Richard B., 1950–
   II. Title.
   QK100.N6L48   1984   581.6′9′097   84–1505
   ISBN 0–8289–0531–2
   ISBN 0–8289–0530–4 (pbk.)

Contents

Preface *xi*
Introduction *1*

Part I Plants That Cause Dermatitis 9
Poison Ivy *13*
Poison Oak *15*
Poison Sumac *15*
Controlling Poison Ivy, Poison Oak and Poison Sumac *16*
Poisonwood *17*
The Mango Tree *18*
The Manchineel Tree *20*
The Cashew Tree *22*
Polyscias *24*
Brazilian Pepper, Florida Holly *24*
Stinging Nettles *25*
Plants Containing Photosensitizers *27*

Part II Hallucinogenic Plants 28
Peyote Cactus *30*
Cannabis, Marijuana *32*
Morning Glory *35*
Thorn Apple, Jimsonweed, Jamestown Weed *37*
Angel's Trumpet, Devil's Trumpet *39*
The Matrimony Vine *40*
Opium Poppy *41*
Atropa belladonna *43*
Black Henbane *44*

Part III Toxic Plants of the Home and Garden 45
LILY FAMILY *46*
Lily-of-the-Valley *48*
Daffodil *49*
Climbing Lily *49*
Autumn Crocus, Meadow Saffron *50*
Asparagus *51*
MUSTARD FAMILY *51*
Castor Bean Plant *52*
Foxglove *54*
BEAN FAMILY *56*

v

Rosary Pea, Precatory Bean, Crabs-Eye, Jequerity, Prayer Bean, Love Bean 57
Lupines 58
Black Locust 59
Cultivated Beans 60
Wisteria 61
THE NIGHTSHADE FAMILY 62
Potato, Irish Potato 63
Woody Nightshade, Climbing Nightshade 63
Deadly Nightshade, Black Nightshade 64
Horse Nettle, Wild Tomato 65
Jerusalem Cherry 66
Jessamine 67
Ground Cherry, Chinese Lanterns, Tomatillo, Cape Gooseberry, Strawberry Tomato 67
Chalice Vine, Trumpet Flower 68
AROID FAMILY 68
Rhubarb, Pie-Plant 71
Ivy, English Ivy 73
Yew, Ground Hemlock 74
Horse Chestnut, Buckeye 76
Cycads 77
Chinaberry Tree, China Tree, Texas Umbrella Tree 78
Sandbox Tree 79
Jatropha 79
Honeysuckles, Elderberries, Coral Berries, Snowberries, Viburnums 80
Tung-Oil Tree, Tung Nut 82
DOGBANE FAMILY 83
Oleander 84
Yellow Oleander, Lucky Nut, Be-Still Tree 85
RHODODENDRON FAMILY 86
Lambkill, Sheep Laurel, Mountain Laurel, Bog Laurel 87
Daphne, Mezereon, Spurge Laurel 88
CHERRY AND APPLE FAMILY 90
Privet 91
Golden Chain 92
Cassava, Manioc, Tapioca, Yuca 93
SPURGE FAMILY 94
Red Spurge, 94
Crown-of-Thorns 94
Pencil Tree Cactus 94
Candelabra Cactus 96
Snow-on-the-Mountain 96

Poinsettia 97
Other Annoying Spurges 97

Part IV Wild Poisonous Plants 98
Celandine Poppy, Rock Poppy 99
Poison Hemlock, Water Hemlock, Fool's Parsley *101*
Pokeweed, Pokeberry, Pigeonberry *104*
Death Camas, Black Snakeroot *106*
Milkweeds *107*
Buttercup, Crowfoot *108*
Baneberry, Doll's Eyes, Coralberry, White Cohosh *110*
Anemone, Wind Flower, Pasqueflower, Thimbleweed *110*
Marsh Marigold, Cowslip *110*
Virgin's Bower *111*
Christmas Rose *111*
Indian Tobacco *111*
Yellow Jessamine, Carolina Jessamine *112*
BITTERSWEET FAMILY *114*
Coyotillo, Tullidora, Capulinicillo, Gallita Bush *115*
Buckthorn, Alder Buckthorn *117*
Delphinium spp., Larkspur *118*
Aconitum spp., Aconite, Monkshood, Wolfsbane *118*
Mistletoe *120*
False Hellebore, Indian Poke, Corn-Lily *121*
Lantana *123*
Golden Dewdrop, *Duranta* *124*
Holly *124*
Tansy *126*
White Snakeroot *126*
May Apple, Mandrake *128*
Moonseed *129*

Part V Poisonous Mushrooms 131
I. DEADLY OR POTENTIALLY DEADLY MUSHROOMS *136*
Deadly *Amanitas* *136*
Death's Angel, Death's Cap *137*
Deadly Little Brown Mushrooms *139*
Deadly *Conocybe* *141*
False Morels *141*
Neogyromitra gigas *143*
Helvella lacunosa *143*
II. MUSHROOMS THAT CAUSE PARASYMPATHETIC STIMULATION *144*
Fly Agaric *144*

Panther *Amanitas* *146*
III. MUSHROOMS THAT CAUSE SWEATING, TEARS, AND SALIVATION *148*
Clitocybes *148*
Inocybes or Fiber-Heads *149*
IV. HALLUCINOGENIC MUSHROOMS *150*
Psilocybes *150*
Bog *Conocybe* *153*
Girdled *Panaeolus* *153*
Gymnopilus spectabilis *153*
V. MUSHROOMS TOXIC ONLY WITH ALCOHOL CONSUMPTION *154*
Inky Caps *154*
VI. MUSHROOMS THAT CAUSE GASTROENTERITIS *155*
Russula emetica *156*
Milky Mushrooms *157*
Jack O'Lantern or False Chanterelle *158*
Hebelomas *159*
Entoloma *160*
Verpa bohemica *161*
Poisonous Tube Fungi, *Boletes* *162*
Paxillus involutus *164*
Cortinarius *165*
Other Poisonous *Amanitas* *165*
Selected References *169–171*
Index *173*

List of Color Plates

These color plates follow page 82.

Plate 1. Stinging Nettles
Plate 2. Marijuana
Plate 3. Morning Glory
Plate 4. Thorn Apple, Jimsonweed
Plate 5. Opium Poppy
Plate 6. Lily-of-the-Valley
Plate 7. Daffodil
Plate 8. Autumn Crocus
Plate 9. Castor Bean Plant
Plate 10. Jack-in-the-Pulpit
Plate 11. Elephant's Ears
Plate 12. Rosary Pea
Plate 13. Lupines
Plate 14. Climbing Nightshade
Plate 15. Yew
Plate 16. Horse Chestnut
Plate 17. Oleander
Plate 18. Mountain Laurel
Plate 19. *Rhododendron* species
Plate 20. Mezereon
Plate 21. Crown-of-Thorns
Plate 22. Poison Hemlock
Plate 23. Pokeweed
Plate 24. Milkweed
Plate 25. Bittersweet
Plate 26. May Apple
Plate 27. Destroying Angel
Plate 28. Destroying Angel
Plate 29. Eastern Fly Agaric
Plate 30. False Morel
Plate 31. *Galerina autumnalis*
Plate 32. False Chanterelle

Preface

THIS book is intended to provide comprehensive, up-to-date coverage for the identification, symptoms and medical treatment of common toxic fungi, wild plants, and common ornamental plants in North America. Most existing field guides to the plants and mushrooms are of dubious value, since they sometimes list frankly dangerous species as either edible or questionably edible. Such books often reflect their authors' experience and unfortunately the same species that is relatively harmless in one part of the country may be rather toxic elsewhere. While there are also a number of books on poisonous and cultivated plants, most are either out of date, very technical, oriented toward animal husbandry, or superficial. Books on herbal medicine have minimized or even ignored the potential hazards associated with using homemade plant medicines. These books also lack clear, current advice on medical treatment. To overcome these deficiencies, this guide incorporates a number of features which make it easy to use by people not trained in the plant sciences.

We have tried to dispel the mysteries of medical and botanical technical jargon by using everyday language that will not require frequent references to a glossary. Each plant and fungus has its key identification features clearly described and is accompanied by an accurate line drawing. Each drawing in turn has labels indicating major identifying characteristics, a feature not offered by other currently available books. There is also a section which describes the usual geographic distribution of the plant or fungus, its normal habitat, and seasonal occurrence. This section is followed by a brief description of the poisons found in the plant and a detailed description of the symptoms of poisoning. This is important because most physicians have little experience or training in dealing with cases of plant poisoning. The symptoms section also includes an evaluation of the damage each poison can cause. Unfortunately, most people tend to overreact to any signs of poisoning, and it is well known that fear and anxiety aggravate any symptoms. This is not to make light of plant poisoning, and should symptoms present themselves after consumption of a plant or fungus, you should call your local poison control center and, if possible, bring a sample of the plant along with the patient in order to get positive identification. Finally, there is a section on first aid and medical treatment, presented in non-technical language.

The section on medical treatment is based on a detailed review of the most recent medical literature. Several physicians have reviewed this material for accuracy. Bring this book with you to the physician's office or hospital, since many doctors have little knowledge of the subject, or their books are many years out of date. If plant poisoning does occur, make sure that your local hospital or physician reports the case to the local poison center, since this data is important and is often lacking.

We are very grateful to Frederick Lovejoy, M.D., Assistant Physician in Chief of Children's Hospital in Boston and director of the Massachusetts Poison Control Center, for his constructive critical review of this book. Despite our research, there is much confusion and controversy regarding the management of plant poisoning and Dr. Lovejoy provided the most up-to-date insights that only an experienced toxicologist and clinician could provide. We are also fortunate to have gotten the critique of the mushroom section by a physician experienced in mushroom toxicology and identification, Dr. Lot B. Page, Chief of Medicine at Newton Wellesley Hospital. All botanical names and descriptions have been reviewed for accuracy by Dr. Elizabeth Shaw, a plant taxonomist at the Harvard University Herbarium.

We were aided in our literature search by the six-year medical students at Boston University and we note our appreciation. The artwork is totally new and provided by two talented artists. Mr. L. Laszlo Meszoly, who teaches at the Harvard Museum of Comparative Zoology, provided the drawings of all of the fungi while Margaret Huong Primack drew the majority of the plants using materials from the Boston University and Harvard University Herbarium Collections. While many photographers contributed color plant portraits, we would like to particularly thank Ms. Frederica Matera, who traveled far and wide to get us specific pictures. Finally, we wish to thank Ms. Shirley Manditch for typing the manuscript.

A FIELD GUIDE TO
*Poisonous Plants
and Mushrooms
of North America*

Introduction

AS plants evolved, so did a vast army of insects and vertebrates with a ravenous appetite for plants. The destructive potential of the animal predators put an enormous survival pressure on plants to develop a variety of countermeasures that would make them less vulnerable to attack. Since plants cannot run away, they have evolved a variety of defensive weapons—powerful chemical poisons, sharp thorns, prickly hairs and irritants. Co-evolving with the plants' defenses, the predators adopted new strategies and tactics requiring new improvements in the plants' own natural defenses. For example, they developed even more potent chemical poisons to make them inedible to munchers.

The development of blatantly poisonous compounds by plants and fungi is extraordinary in the variety of toxins that they produce. These compounds are chemically very diverse and include powerful substances that affect heart muscle and blood pressure, smooth muscle relaxants, cyanides that block cell respiration, cell poisons that inhibit protein synthesis, hormone-like compounds, hallucinogenic chemicals, irritants, blistering agents, photosensitizers and potent allergens. Some act rapidly, causing instant irritation, nausea, vomiting and diarrhea, while others are more insidious, producing deadly delayed reactions. While the development of these potent and sophisticated chemical defenses has helped plants and fungi avoid being eaten, these poisons have also caused deaths, pain, itching and a variety of ills to people who have either eaten or come in contact with them.

POISONOUS PLANTS: HOW MUCH OF A PROBLEM?

While there are some 7,000 plants and fungi that produce or contain toxic substances, only a few are really very dangerous. According to the Food and Drug Administration's National Clearing House for Poison Control Centers, there were only 7,710 cases of exposure to plant poisons recorded in 1975. Of these victims, 1,990 reported symptoms, 186 were hospitalized and 3 died. However, Dr. Guy Hartman, a California pediatrician familiar with plant poisoning, reported 2,845 cases of plant poisoning just in his area in 1976. According to Dr. Hartman, wild plants cause the majority of summer exposures and poisonings, while house plants are the usual culprits during the winter. Of his 2,845 cases, 89 were caused by hallucinogenic plants, toxic mushrooms accounted for 419 cases, castor beans for 31 cases, oleander for 124 cases and the remainder were attributed to other plants.

Given the fact that diagnosis of plant poisoning is very difficult and reporting of plant poisoning by hospitals and physicians is not mandated by law, our insight into the dimensions of the problem is at best sketchy. Dr. Hartman estimates that there are probably 40,000 cases of exposure to plant poisons per year in the United States, but that even this estimate may be two or three times too low.

WHO IS AT RISK?

The number of people gardening and using ornamental plants for home decoration has quadrupled in the last two decades, and, according to the Department of Agriculture, 48 percent of the population (over one hundred million people) are somehow involved. Visits to the National Parks, National Forests and State Parks have more than doubled as outdoor recreations (hiking, backpacking, hunting, fishing, bird watching) have become popular escapes. Ever-increasing numbers of people are gathering wild plants in search of new gastronomic natural treats, bringing into jeopardy another segment of the population. Some of these wild harvests involve misidentified plants and can cause a most unhappy or even deadly experience. The number of people practicing herbal medicine (a tradition that goes back to before the time of Christ) is also on the rise. People seeking natural products (roots, leaves, and bark) to make their concoctions and potions can err and experience mild to severe poisoning. Another group of outdoors people who use wild plants, sometimes with considerable risk to themselves, are those people looking for a natural high from smoking or eating plants which contain hallucinogens, although the greatest risk here comes when such a person stumbles across a hidden marijuana plot guarded by a trigger-happy protector of his crop. There have been several deaths due to such accidental encounters.

Ornamental house and garden plants in and around the home pose a particular threat to children. According to Dr. Lovejoy, Director of the Massachusetts Poison Control Center, even the ever-popular *Philodendron*, although not particularly dangerous, is eaten by 45 Massachusetts children each month. Some of these children exhibited symptoms requiring medical attention. There are other common house and garden ornamentals that pack considerably more potent toxins, and these include a number of bulbs that are often around within children's reach prior to planting.

The family vegetable garden is another potential hazard to children, with rhubarb leaves and the green parts of potato and tomato plants, if ingested, causing symptoms and even deaths. Finally, the family fruit trees, such as apricot, peach, plum, cherry, and even apple, produce seeds with measurable amounts of cyanide, one of the most deadly inhibitors of cellular respiration.

With many of these poisonous plants, the primary population of victims are young children. Children will put almost anything in their mouths, and anyone who has had to keep watch over a two- or three-year-old can appreciate his constant exploratory behavior and endless energy. While the number of children ingesting potentially toxic plants has increased, the reality is that a few munches on a leaf will usually not put the child in jeopardy. While there is no need for panic, there is reason to exert a little good common sense. Poisonous ornamental house and garden plants such as oleander and castor bean plants should be removed from homes that have young children. Even less toxic plants should be placed out of reach, and the poisonous attractive berries found on yews and pokeweed should be removed. An ounce of caution is worth a pound of cure, and this complete guide to poisonous plants of field, home, and garden will allow you to poison-proof your children's living spaces. To put the hazard in perspective, far more children are poisoned by insecticides, household cleaners and medicines that are carelessly left within the reach of toddlers than by natural plant poisons, so when poison-proofing your living area, make sure that you put other poisonous material in sealed containers, out of reach or in locked cabinets.

While prevention is most important, some people will be exposed to plant poisons, that is, eat toxic plants but not get symptoms. A small number will actually be poisoned, that is, experience symptoms requiring treatment. In order to manage acute poisoning successfully we have set up a list of common-sense rules that should be followed.

SOME COMMON-SENSE RULES TO DEAL WITH THE PROBLEM OF POISONOUS PLANTS

1. If a suspected toxic plant has been eaten, the first thing to do is to call the local poison control center as soon as possible.

These centers are staffed by trained people familiar with poison exposure and you should follow their instructions carefully. Give the poison control center the following information: (1) the age and weight of the victim, (2) the time of suspected ingestion, (3) describe any symptoms as accurately as possible, (4) give them the name of the plant, preferably its scientific name as well as its common name, (5) estimate how much of the toxin has been ingested. If possible, bring any uneaten parts of the plant with you to the emergency room or physician's office. *Drowsy, convulsing, or unconscious poisoning victims should be given nothing orally and be rushed to the nearest hospital.*

2. Don't panic—most plant poisonings are relatively mild, and your overt overreaction can amplify the symptoms, since children often pick up parents' feelings.
3. Know which plants are poisonous and what their toxic potentials are.
4. Remove the more dangerous garden and house ornamental plants (oleander, castor bean, Jerusalem cherry, dumb canes, etc.) if you have small children.
5. Some items of jewelry are made from seeds (such as rosary pea and castor bean) that contain potentially deadly toxins; some handcrafted jewelry produced in the Tropics and Asia is coated with a lacquer resin causing skin irritation in sensitive individuals; get rid of this type of jewelry or keep it out of reach of children.
6. Don't eat any raw wild vegetables (particularly bulbs) in salads unless you have absolutely positive identification.
7. Don't chew on unfamiliar leaves, grasses or stems of wild plants, and train your children not to eat unknown plants.
8. Wild mushrooms can be lethal; unless expert identification has been made, avoid them and stick to mushrooms bought at the supermarket.
9. Do not eat or touch plants that ooze milky or colored plant juices. Some are poisonous, some cause dermatitis and some irritate moist membranes of the eyes, nose and mouth (exceptions: lettuce stems and dandelions).
10. Avoid *unknown* red or blue-purple berries, as many brightly colored fruits are toxic. Fruits that are eaten by birds are not necessarily safe for human consumption.
11. Don't eat unknown wild seeds, because once the seed coat is cracked, you can be exposed to accumulated toxins within the seed.
12. Do not eat bulbs, roots or tubers of unknown plants. Many, such as Amaryllis or Narcissus, are toxic, particularly if they have an onion-like appearance.

13. Don't eat fruits that are three-lobed (spurge, horse chestnut, etc.). They are potentially deadly.
14. If you enjoy gathering wild plant foods, cook them with two changes of water, since some of the poisons may be water-soluble and are inactivated by heat. However, be warned: Some poisons are not destroyed by cooking.
15. If you feel that you must experiment with wild plants as foods, don't over-indulge; try small samples and wait before trying the plant again.

WHAT TO DO IF A TOXIC PLANT IS EATEN

Whenever part of a known or suspected toxic plant is eaten, the most important first step is to promptly remove the suspected material from the victim's stomach. The longer it stays in the stomach, the greater the amount of absorption of the toxin and the greater the degree of irritation to the gastrointestinal tract. Many toxic plants cause emesis or vomiting, and if this occurs the stomach contents are emptied, usually along with the offending plant parts. If the victim does not spontaneously regurgitate, call the poison center and they will advise you. Usually you will be advised to induce vomiting *as soon as possible* and save the vomit to bring to the emergency room, since it may contain plant parts that are needed for identification.

WHAT TO DO TO INDUCE VOMITING

Many of the older handbooks, texts, and articles give contradictory advice about how to treat a patient who has ingested toxic plant material. To resolve this controversy, we follow the guidelines of Drs. Easom and Lovejoy of the Massachusetts Poison Control Center. The first step is to empty the contents of the stomach. *Do not* induce gagging with a finger or use solutions of salt, mustard, egg white, or copper sulfate; they can do harm. The so-called "universal antidote" suggested by many older books is worthless; don't bother with it. The preferred method is to give the patient Syrup of Ipecac. The usual dose is 30cc (2 tablespoons) for adults and 15cc for children. This treatment may be repeated once in about 20 minutes but do not exceed a total dose of 60cc in adolescents and adults or 30cc in children. Vomiting should occur in 14 to 24 minutes in over 90 percent of patients. Parents of young children might be wise to keep a bottle of Syrup of Ipecac in their medicine cabinets. After the victim has vomited give him lots of water. IF A POISONING VICTIM IS DROWSY DO NOT INDUCE VOMITING SINCE HE/SHE MIGHT INHALE OR ASPIRATE THE VOMIT.

How Do You Lessen Absorption of Ingested Plant Toxins?

Once the victim reaches medical help, they will administer activated charcoal (Norit AR) in water (about 20 to 50 gms in eight to 16 ounces of water) orally. This procedure will reduce absorption into the body by absorbing much of the toxin on its highly absorbitive surface. Children should be given 1 gm per kgm body weight in eight to 16 ounces of water. The charcoal will not confer any benefit in cases involving cyanide-containing plant poisons. Another means of decreasing absorption of the toxin into the bloodstream is to clear the ingested plant material through the bowel as rapidly as possible. Some poisonous plant toxins are irritants which cause diarrhea; others don't. Give the victim a saline laxative such as magnesium sulfate to accelerate transit through and evacuation of the bowel. Do not use oil-based laxatives.

What Do You Do if Plants Containing Cyanide are Eaten?

A number of plants contain cyanogenic toxins, which are deadly inhibitors of the vital iron-containing respiratory enzyme cytochrome oxidase. To counter the cyanide effect, use the Cyanide Antidote Package® (Lilly Co.), which contains sodium nitrite, amyl nitrite, and sodium thiosulfate. These compounds act to bind or inactivate some of the absorbed cyanide. The Cyanide Antidote Package® should be used with caution. Check the expiration date and follow the dose carefully, particularly when children are the victims of poisoning. Oxygen therapy reverses cyanide binding to respiratory enzymes and provides more oxygen to tissues and it also increases the effectiveness of the cyanide antidote kit. A new promising treatment for cyanide poisoning is Vitamin B_{12a} in a concentration 50 times the amount of absorbed cyanide.

What Do You Do if Plants Containing Digitalis-Type Alkaloids are Eaten?

Ingestion of plants containing digitalis-type alkaloids should be treated with prompt emptying of the stomach followed by activated charcoal and laxatives repeated at four-hour intervals until the stool contains charcoal. Serum potassium levels should be monitored and potassium should only be given to hypokalemic (low-blood potassium) patients. Cardiac effects require specific treatment: Phenytoin for ventricular arrhythmias (irregular heartbeat), Atropine for sinus bradycardia (slow beat), Lidocaine and Propranolol for ventricular tachycardia (fast beat) and electrical pacing for heart block (asynchronous beat).

WHAT DO YOU DO IF IRRITATING SAP IS ENCOUNTERED?

Plant irritants in the mouth or eye should be removed by repeated flushings of cool tap water (no other mouth- or eyewash should be used). Continue flushing for about 20 minutes. If the eye is involved, the victim should be seen by a medical eye specialist (ophthalmologist).

ORGANIZATION OF THIS BOOK

This book identifies all of the commonly encountered poisonous plants and fungi of North America. Certain species are omitted that are of minor importance due to infrequency of occurrence or relative mildness of the symptoms. This book uses English measurements of length throughout due to their common use in North America. The book is divided into five sections.

1. The first section deals with plants that cause itchy contact dermatitis—common plants such as poison ivy, poison oak, poison sumac, and their kin; plants that afflict up to 60 percent of the population, particularly those who enjoy outdoor activities.
2. The second section identifies hallucinogenic plants that are grown or sought by people who use them for recreational purposes, sometimes with disastrous results. We have not included plant-derived drugs that are imported, such as cocaine and nutmeg.
3. The third section covers plants commonly found in or around the house. These ornamentals, though handsome, can be deadly, particularly to children, who have a tendency to put all manner of things in their mouths.
4. The fourth section deals with common poisonous wild plants, some of which look like edible species. Among these are some very poisonous plants.
5. The last section deals with poisonous mushrooms, or toadstools. While most of these produce only moderate to severe

gastrointestinal upsets similar to those caused by food poisoning agents, bacteria, and viruses, some are potentially deadly and responsible for a number of deaths each year. Some fungi have no effects by themselves, but if you consume alcohol after eating them you become quite sick; some cause sweating, salivation, lacrimation (watery eyes) and other autonomic effects; while others are sought after because they contain potent hallucinogenic substances.

We consulted over 400 references in researching this book; however, the book is not intended as a scholarly reference, but as a practical manual for the general public and physicians. We have included a list of selected references for those readers wishing to dig deeper into the subject.

PART I
Plants That Cause Dermatitis

AS an escape from society's technology and complications, many people seek the solace and comfort of nature's beauty and peace. However, with all benefits there are risks, and one of these is the risk from plants that cause allergic contact dermatitis. By far the most common scourge of outdoors people are members of the widely distributed Anacardiaceae family of plants, which include poison ivy, poison oak, poison sumac, and several other insidious villains which will be described in the following pages.

The number of people coming in contact with poison ivy and other members of the Anacardiaceae family is quite high—many investigators having shown 40 to 75 percent of North Americans to be sensitive to the urishol toxins. Therefore, an important factor in the management of plant contact dermatitis, as in that of other types of allergic contact dermatitis, would appear to be in the recognition and avoidance of the causative agent.

Poison ivy and poison oak are the principal culprits in contact dermatitis. Western poison oak, *Toxicodendron diversilobum*, is more prominent on the West Coast than elsewhere, while Eastern poison oak, *T. toxicarium*, is limited to the Atlantic and Gulf coastal plains. Poison ivy, *Toxicodendron radicans*, occurs throughout Canada and the United States except in the extreme Southwest, and in the coastal Pacific Northwest. In the older literature, members of the genus *Toxicodendron* were included in the genus *Rhus*, but this name is now restricted to the sumacs, a group to which these poisonous species are closely related.

Toxins: Though there are physical differences between toxic members of the Anacardiaceae and differences in their habitat, the allergic contact dermatitis produced by their toxins is the same, and cross-reactions may occur with the following related plants and substances:

1. A furniture and jewelry lacquer obtained from the Japanese lacquer tree *(Toxicodendron verniciferum)* also contains urishol.

10 Poisonous Plants and Mushrooms of North America

2. The oil from the shell of the cashew nut *(Anacardium occidentale)* and the juices of the raw cashew nut contain a related potent sensitizer.
3. The rind of the mango fruit *(Mangifera indica)* contains an allergen similar to the toxic substance of poison ivy.

There are a number of predisposing factors influencing a person's susceptibility to urishols, but in all probability sensitivity to particular substances is genetically controlled. Some individuals will never contract poison ivy dermatitis because they are resistant to sensitization. There are also racial differences in sensitization to poison ivy, with blacks generally more resistant than Caucasians. Age has little influence on sensitization, although after the age of 70 sensitivity decreases. Even infants can be sensitized to poison ivy. Most allergenic substances encountered in daily life cause sensitization only after repeated contact. The time between first contact and the onset of the sensitization depends partly on the sensitizer and the conditions of exposure, partly on constitutional factors and partly on coincidental lowering of the resistance of the skin.

The interval between the start of sensitization and its completion is called the "period of incubation" (the time between the sensitizing exposure and appearance of the symptoms). Present evidence indicates that the sensitizing processes are complete within five days. Even after the antibodies (IgE) are formed in response to the allergen, a further 24 to 48 hours will pass before the development of symptoms. Once acquired, contact sensitivity tends to persist, although the degree of sensitivity may decline with time. If the last attack occurred within five years, symptoms will usually erupt within two days, but if the last attack was six to 20 years ago, symptoms will not appear until five to seven days after exposure.

The agent responsible for the allergic reactions of poison ivy contact dermatitis is a mixture of oleoresins commonly referred to as urishols. Dermatitis usually begins with itching and redness of the skin (sometimes as streaks) or papules (raised itchy bumps), vesicles (small fluid-filled blisters) and bullae (large fluid-filled blisters), often in linear arrangement. Contrary to popular myth, the fluid from the blisters does not produce antibodies or cause spread of the eruption. Poison ivy dermatitis can come only from direct contact with oleoresin-contaminated animals, clothing, tools, aerolized droplets in the smoke of burning plants or direct contact with bruised portions of the plant. The oleoresin may remain under a person's fingernails, and, if it is not removed, can, for several days, spread the eruption to other parts of the body and even to other individuals.

11 Plants That Cause Dermatitis

Large vesicles (blisters) or bullae should be opened under sterile conditions and with preservation of the lesion (leaving the overlying skin) if possible. Swelling of the eyelids can be treated with wet dressings of cold boric acid. On the face, lotions should usually be avoided in acute allergic dermatitis, because they tend to cake, making the skin uncomfortably stiff. Applications of wet dressings followed by zinc oxide ointment or petrolatum are better tolerated. After vesicles or bullae are opened, baths with potassium permanganate (one teaspoonful of potassium permanganate crystals dissolved in a tub of lukewarm water) are beneficial, both for their quick-drying effect and for their efficacy against secondary infection.

Complications: Among these is the development of secondary bacterial infection. The contents of vesicles and bullae become cloudy, and pustules of impetigo develop near the sites affected by the dermatitis. If this infection is superficial and not excessively inflammatory, local treatment will suffice. Removal of infected debris by washing with soap or soap substitute containing hexachlorophene, followed by treatment (three times daily) with an ointment containing neomycin and bacitracin, should produce improvement within 48 hours. Sweat retention due to inflammatory blockage of sweat ducts is not serious unless the sweat glands are not functioning at all; then the only satisfactory means of dealing with the problem is to reduce the skin temperature.

Areas of skin which are damaged by contact dermatitis may develop a prolonged "secondary dermatitis," particularly in moist, thin-skinned areas such as the anogenital region, or in areas previously damaged.

Several peculiar cases of poison ivy dermatitis have been described in which patients, in addition to having typical symptoms, developed black lacquer-like deposits that could not be removed with soap and water.

Prevention and First Aid: The best way to prevent plant contact dermatitis is the complete avoidance of plant sap from the leaves, stem, or any plant part, which is difficult in the outdoors. While clothing (long pants and long sleeves) offers protection, contaminated clothes must be washed in hot, soapy water. Topical measures (barrier creams) are not effective in the prevention of poison ivy dermatitis. Washing with hot, soapy water immediately after exposure will remove the urishol, but since this substance enters the skin rapidly, the oil must be totally removed within *ten minutes* of exposure if symptoms are to be prevented.

Medical Treatment: (Based on procedures recommended by Dr. L. E. Milliken, *Hospital Practice*, July 1980). For mild cases (limited area of inflammation—less than 10 percent of the body), use cold soaks with tap water or preferably Burow's solution® soaks applied for 20 minutes, four times a day. Non-prescription topical corticosteroid (0.5%) creams or, better yet, physician-prescribed (1% to 3%) creams should provide relief sooner than just soaks. If the little blisters are leaking fluid, use loose Kerlix® over the oozing area and hold it in place with Coban® tape. *Soak* the bandage off at each soak cycle. Moderate cases (10 to 30 percent of the body, but not including face, hands or groin): Burow's® soaks four times a day to dry the blisters. Use corticosterone cream four times a day to ease itch and inflammation. Be alert for secondary infection. Severe cases (hands, face or groin involvement, more than 30 percent of the body): See a physician, who will give systemic (oral) corticosteroids such as prednisone (0.5 to 0.7 mg/kg of body wt/day) for four days. Decrease by 10 mg every four days down to 20 mg/day. Alternate days, 20 mg for one week. Usually the recovery is complete in two or three weeks.

Topical (applied onto the skin) preparations containing benzocaine, zirconium or antihistamines should be avoided, because these substances sometimes act as sensitizers. The old standbys such as calamine lotion and other soothing concoctions are of limited value.

POLLEN-CAUSED ALLERGIC CONTACT DERMATITIS

We are all familiar with the effects of plant pollens on the respiratory system, but many pollens contain another antigen which is an oleoresin capable of producing contact dermatitis on exposed parts of the skin. Usually the response is less severe than that of poison ivy because pollen grains are microscopic and contain less oleoresin. The most common villain is our old enemy ragweed, and dermatitis can even be caused by contact with the leaves and stems of this common annoying weed. Other members of this family can also cause allergic contact dermatitis. These include chrysanthemums, daisies and asters. Some people also get dermatitis from some tree pollens, e.g., box elder, poplars and maples.

PLANT IRRITANTS THAT CAUSE DERMATITIS

A number of plants are capable of damaging the skin and producing itchy inflammation because they contain acrid irritant chemicals in their juices or have spines or penetrating hairs capable of piercing the skin barrier. The irritants are not allergens; hence there is no sensitization process involved. There are, however, individual differences in susceptibility, such as thickness of the skin; thus children, with their delicate skin, are the most frequently affected. Among the plants that contain potent chemical irritants are: buttercups (see p. 108), spurges (see p. 94), some mustard plants and radishes, gourds, and even cucumber plants.

Plants, in the course of their evolution, developed mechanical protective devices as well as chemical warfare capability. To discourage browsing munchers, they developed quite an arsenal: sharp thorns and spines or slender, penetrating hairs and bristles, and sometimes combined these with irritant chemicals. The symptoms they produce range from sharp localized pain at the site of penetration to itchy blisters. The larger structures, such as thorns, can be seen imbedded in the skin and can be removed with tweezers, but the fragile irritating bristles, hairs (trichomes) and barbs break off in the skin, and the only way to get rid of many of them is to use peeling ointments (e.g., salicylic acid). Always be alert for complications, such as bacterial or fungal infection, that may accompany mechanical irritants.

POISON IVY

Of all members of the family Anacardiaceae, one stands head and shoulders above the rest as the main source of human discomfort. This common malicious plant, *Toxicodendron radicans* (formerly *Rhus radicans*), commonly called poison ivy, poison vine, or markweed, is a hardy, prolific, innocuous-looking plant that is widespread throughout much of North America. It can thrive in a variety of habitats—the edges of sand beaches, along stone walls and fences, in the woods in light shade, in parks and even in the family backyard. One reason for its wide distribution is that its berries are eaten by about sixty species of birds and the seeds are then disseminated in the birds' droppings. This deciduous plant can grow as a woody climbing vine or shrub, a ground cover, or a low thicket. The leaves are divided into 3 notched or lobed leaflets, with the center leaflet often on a longer stalk. There are several varieties that look somewhat different than the eastern race described above. In the central states, the leaflets have toothlike notches. In Oklahoma and Texas, the terminal or central leaflet has a deep lobe on either side, while the lateral leaflets have only a single lobe; in the Rio Grande region the leaflets, though deeply cut, are round-lobed, somewhat in the shape of clubs on playing cards. All species can exhibit considerable variation in leaf size, color and shape, depending on the amount of

14 Poisonous Plants and Mushrooms of North America

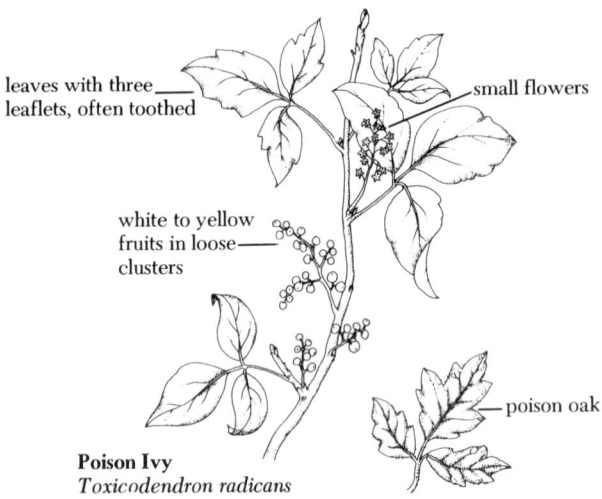

Poison Ivy
Toxicodendron radicans

available light. In certain forms, the two lateral leaflets are lost, and the remaining central leaflet looks like a simple leaf. In brightly lighted areas the leaves may be larger and have a lustrous sheen, while in shaded areas the leaves are thinner, dull and narrower. Along the edges of dirt roads plants may become dust-covered and difficult to recognize, but the dust doesn't make them any less potent. Vines may reach enormous lengths—up to 75 feet long—and climb to considerable heights on trees. They can be recognized in the vine form by the reddish mat of hairs which grows onto the host tree's trunk. In its shrub-like form it may assume the appearance of a small shrublet or hardy bush. The small yellow-green flowers have five petals, five stamens and a solitary pistil. Late in the summer, particularly once the leaves have fallen, you can see its globular, pea-sized fruits, which are white, tan, or yellowish and contain a single seed.

In the fall the leaves become brilliantly colored—reds, yellows, oranges—and many is the naive hiker who has brought home a colorful, itch-producing bouquet of fall foliage. The fallen *T. radicans* leaves sometimes end up in the leaf pile, and in the days when burning leaves was a fall ritual, there would be an outbreak of poison ivy dermatitis caused by droplets of urishol carried in the smoke. Smoke from burning poison ivy leaves and other plant parts can cause dangerous symptoms if the smoke gets into the lungs and eyes. (see cover, mid left)

Distribution and Habitat: Throughout most of the United States and southern Canada except in the Pacific Northwest and extreme

15 Plants That Cause Dermatitis

Southwest; often in dense stands along roadsides and rivers and on sand dunes. It will not grow in dense shade, but will thrive where open lighted spaces have been created by fallen trees.

Toxins and Symptoms: See pp. 9 and 10.

First Aid and Medical Treatment: See p. 11.

Poison Oak

Similar to poison ivy are the two species of poison oak, the western variety, *T. diversilobum*, and the eastern variety, *T. toxicarium*. Both of these cousins are closely enough related that they may interbreed with poison ivy. The eastern poison oak is very similar to poison ivy in leaf pattern, but its leaves have velvety hairs on their lower surfaces. The leaves tend to have more rounded tips, and the fruit is often hairy (see drawing for comparison). The western poison oak is also similar in appearance to poison ivy, but the leaflets are larger and the leaf stalks longer. In the fall, poison oak leaves turn deep red. Both species of poison oak produce clusters of small yellow-green flowers, similar to those of poison ivy, that develop into round, white or cream-colored, one-seed fruits that hang in clusters.

Distribution and Habitat: Western poison oak ranges from southern British Columbia southward to northern Baja California and into the Cascades and western Sierra Nevadas to the Mojave Desert. Eastern poison oak inhabits pine savannas of the southern Atlantic and Gulf State coastal plains. Its geographical range overlaps that of poison ivy, and sometimes both can be found together. Poison oak is very adaptable and grows well in poor or sandy soils.

Poison Sumac

Poison sumac, swamp sumac, and poison elder are all common names for *Toxicodendron vernix*, another of the trio of urishol-containing pests of the United States. This tall (up to 15 feet in height) shrub or sparsely branched low tree has compound leaves, with seven to eleven green, smooth-margined leaflets. The stalk, containing many small whitish flowers, emerges from along the leafy branch. The flowers later form a cluster of round, whitish, single-seeded glossy fruits; the non-poisonous sumacs form bunches of red fruits at the top of an erect stem. All parts of the plant contain itch-producing oleoresins, and because of their height are a bane to outdoors people because the leaves frequently brush against the face. In the fall, the leaves turn bright red. The poi-

16 Poisonous Plants and Mushrooms of North America

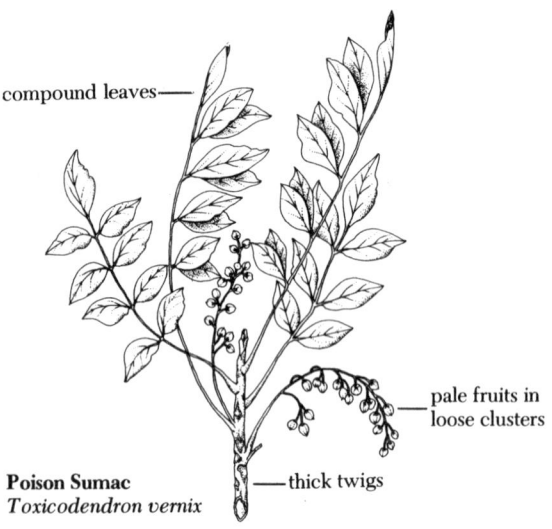

Poison Sumac
Toxicodendron vernix

— compound leaves
— pale fruits in loose clusters
— thick twigs

sonous species might be confused with some of the non-toxic sumacs common to drier habitats.

Distribution and Habitat: Poison sumac is found predominantly east of the Mississippi River, ranging from southern Quebec Province in the north all the way to southern Florida. It is abundant in wet places, swamps, edges of ponds, bogs, and streams.

Toxins and Symptoms: Like all urishol-containing plants, its leaves and other parts when damaged ooze a colorless, gummy sap that contains the oleoresin which produces allergic contact dermatitis. The symptoms are the same as with other plants of this family (see pp. 9 and 10).

First Aid and Medical Treatment: See p. 11.

CONTROLLING POISON IVY, POISON OAK AND POISON SUMAC

In frequently travelled areas around the home, poison ivy and its cousins may be killed with a number of commercially available herbicidal sprays. Spraying should be done on hot, sunny days when there is not much probability of rain for 24 hours. Most of these sprays contain chemicals potentially toxic to humans; hence, be sure to read the instructions carefully and don't leave the herbicide within reach of small children. Ammonium sulfate spraying will also kill poison ivy and other broad-leaved weeds and plants while sparing the grass. Aerosol spray cans specific for poison ivy are commercially available but expensive. Chopping and digging out plants should be done with great caution to avoid contacting any of the sap, either directly or indirectly from gloves, clothing, or tools. Burning plants should be avoided as the poison is carried by smoke, potentially resulting in lung damage if inhaled.

POISONWOOD

Poisonwood, *Metopium toxiferum,* is another member of the large Anacardiaceae family, which includes a number of plant villains that cause contact dermatitis. This large tree can reach heights of up to 40 feet, with trunk diameters of 20 to 25 inches. The trunk has a thin outer layer of flaky brown bark and an orange inner bark layer which, if damaged, exudes a rather sticky sap which blackens on exposure to air. In damaged trees, one often sees streaks and patches of black on the outer bark and spots of black on the leaves. The leaves are six to ten inches long, and each has five to seven broad oval-shaped, green, glossy leaflets (two to four inches long). As these leaflets mature, they turn to a dull, dark green and have a leathery quality. The tree produces sprays of small, five-petaled yellow-green flowers. The flowers are

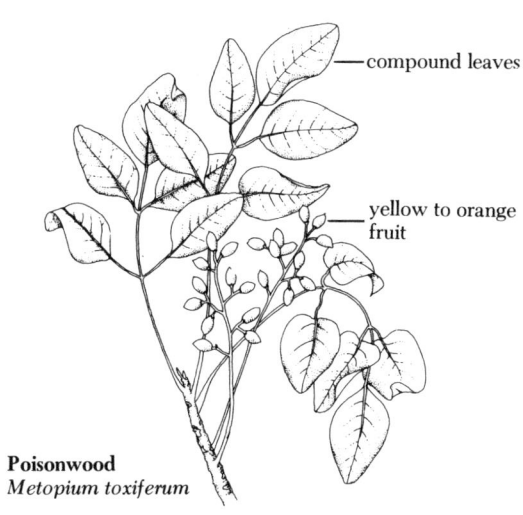

Poisonwood
Metopium toxiferum

compound leaves

yellow to orange fruit

followed by dropping clusters of oval fruits, each a half to three-quarters of an inch in diameter. The fruit is fleshy and contains a single brown seed about one-quarter of an inch long. All parts of the tree contain oleoresins which can cause moderate to severe dermatitis.

Distribution and Habitat: In the continental United States, poisonwood is found only in Florida's southern parts, in pinelands and hammocks from the east coast of Martin County southward. It is also quite common in the Caribbean Islands and Central America, often found in the company of pines and oaks. It can even be found in developed areas, since its seeds are transported both by bird droppings and tropical storms.

Toxins and Symptoms: The toxin from poisonwood, while similar to that of poison ivy, is apparently more potent and produces more severe reactions, even when contacted indirectly. The poison is in the plant resins in all parts of the plant, including the leaves, fruits, and roots. If handling this plant in any way, gloves should be worn, and they should be washed immediately after use, since even indirect contact with this urishol toxin can have an effect. Burning poisonwood produces aerosolized droplets in the smoke, and these can cause temporary respiratory problems and even temporary blindness. The symptoms are exactly like those caused by poison ivy (see pp. 9 and 10), and, depending on the individual's sensitivity, may range from very mild to severe. In severe cases, the poisoning victim may have a fever along with the other symptoms. Individuals with known sensitivity to any of the other plants that cause contact dermatitis should be particularly careful, since cross-sensitivity always occurs. Immunization against these toxins is usually ineffective.

First Aid and Medical Treatment: See p. 11. It is the same for all of the Anacardiaceae.

THE MANGO TREE

Another member of the Anacardiaceae family, a close relative of the cashew nut, is the mango tree (*Mangifera indica*). This handsome, large evergreen tree is native to India and has been widely introduced throughout tropical Central America and Florida because of its tasty fruit. These trees may reach heights of 90 feet and have a canopy spread of as much as 125 feet. The long, rigid, glossy, deep green leaves are six to 16 inches in length. The tasty and valuable pear-sized to cantaloupe-sized fruit is usually green, but may redden on ripening. In cultivated varieties, the fruit's flesh becomes soft and juicy. An average mango contains about

19 Plants That Cause Dermatitis

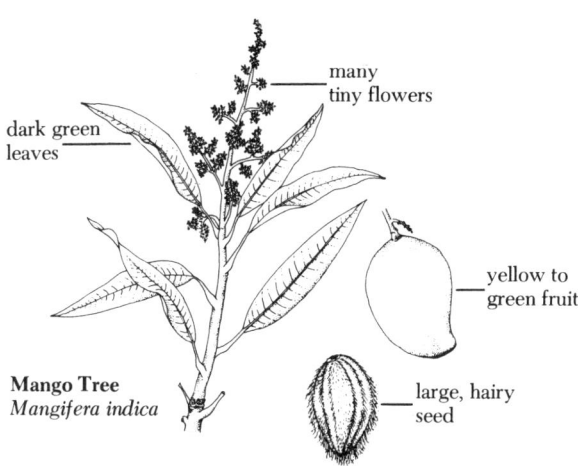

Mango Tree
Mangifera indica

75 calories of fructose and significant amounts of vitamins A and C. Mangoes are eaten fresh or made into preserves; unripe ones can be used to make a hearty relish. They all have a single large, flattened white seed in the middle. All parts of the tree except the flesh of the ripe fruit contain a urishol-like toxin. The clinical picture of mango contact dermatitis seems to be less severe than with poison ivy, poisonwood and cashew urishols.

Distribution and Habitat: Tropical and subtropical North America.

Toxins and Symptoms: The active toxic principle in the mango is called cardol, but is a urishol. It is found in the leaves, stems, bark and pericarp of the fruit but not in the flesh or juice of the fruit. Damaged leaves often drip the toxic resin onto the peel of the fruit, contaminating it and increasing the probability of contact dermatitis. In Hawaii, mangoes are the major cause of dermatitis, and the most usual site of irritation is around the mouth. As is the case with other urishol-like toxins, cross-sensitization occurs, and people with a previous history with other members of this itchy family would be well advised to be cautious around mangoes. The overall symptoms are identical to those described for poison ivy (pp. 9 and 10). However, in severe cases (which are fortunately few in number), anaphylactic shock also occurs due to an exaggerated allergic response.

First Aid and Medical Treatment: Same as for poison ivy, see p. 11.

THE MANCHINEEL TREE (fruits) dangerous ☠

Hippomane mancinella, commonly known as the manchineel tree or poison guava, is one of five species of the genus *Hippomane* and is a member of the poisonous spurge family Euphorbiaceae. A milky white sap is found in all plant parts. It is a handsome, round-crowned tree with dropping branches in a weeping growth form. This tree usually attains heights of 20 feet, though occasionally some may reach up to 40 feet in height and have two-foot-thick trunks. In young trees the pale brown bark is smooth, while on older trees this bark forms thick scales. The leaves are smooth, alternate and broadly ovate, about four inches in length, with a long leaf stalk. The edges of the leaves are finely toothed and the leaves themselves have a lustrous, leathery appearance. The flowers of this tree are small and greenish and arranged in a stiff spike. The sweet-scented apple-like fruit of the manchineel is single or paired, first green, then becoming yellow-to-reddish on maturity. The fleshy manchineel fruit is crabapple-sized and contains an irregular stone that encloses several seeds.

The attractiveness of the fruit as well as its resemblance to the crabapple have caused the poisoning of Spanish conquistadors, shipwrecked sailors and present-day tourists who have eaten them, in some cases with fatal results. The manchineel, despite its poisonous properties, has actually been cultivated in some regions because it is an effective windbreak. The wood from this tree takes a good polish and has been used to make furniture, but it must be scorched first to prevent skin irritation.

Distribution and Habitat: The manchineel tree is generally found just inland from the low tropical mangrove shoreline, on slightly raised sand hammocks. It is often found in the company of some evergreen salt-tolerant shrubs and low trees, and several other deciduous species.

The tree is confined to the coastal zone from the extreme southern end of Florida and the Florida Keys to the West Indies and Central America. Today, *Hippomane mancinella* is rare in the United States except in the Cape Sable region of the Everglades National Park.

Toxins: Different parts of the manchineel tree have different toxic effects. The main toxin is the indole alkaloid physostigmine, which is found in the milky latex. The plant is so poisonous that smoke from its burning wood irritates the eyes, and the latex from the leaves and bark can cause skin swelling. Small wonder that the Carib Indians used the milky sap to poison their arrows! Some consider this tree to be one of the most poisonous American plants.

21 Plants That Cause Dermatitis

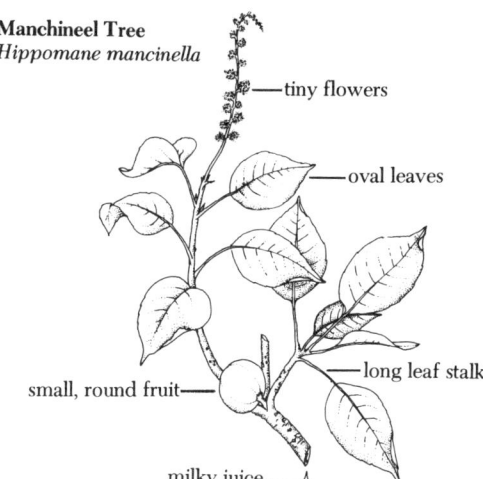

Manchineel Tree
Hippomane mancinella
— tiny flowers
— oval leaves
— long leaf stalk
small, round fruit —
milky juice —

There are accounts of deaths occuring from the effects of the poison when people took shelter from the rain by standing beneath these trees. While this may be somewhat of an exaggeration for the average person, sensitive individuals may get such a severe reaction. The fruit of the manchineel is highly poisonous if ingested.

Symptoms: As is the case with most plants that cause poisoning by contact, the resulting dermatitis is limited almost entirely to man and appears to be rare in domestic and other animals. *Hippomane mancinella* can give one of the worst cases of dermatitis in the country; it abounds in a white milky sap that is highly poisonous. Although not all persons are susceptible and many experience symptoms to varying degrees, poisoning generally consists of either a minor and temporary irritation of the skin beginning one day after exposure or painful irritation and inflammation, with the formation of vesicles or blisters. Temporary blindness or (rarely) permanent blindness is a danger if the eyes are rubbed accidently with latex or smoke gets into the eyes.

Dermatitis resulting from manchineel poisoning generally lasts from a few days to several weeks, depending on the severity of the exposure and the susceptibility of the individual.

The fruit of *Hippomane mancinella* is highly toxic if eaten, and after ingestion it causes abdominal pain, bleeding of the digestive tract, vomiting, and bloody diarrhea. Deaths have occurred. Back in the 18th century, a company of British soldiers sent out as a garrison at St. Kitts in the Virgin Islands, discovered what they thought were edible fruits and bit into them. Unfortunately, they were the fruits of the manchineel tree, and all of the men became dangerously ill and several died in terrible agony. Everyone in the West Indies is familiar with these trees, so instances of ill effects are rare despite their abundance in some areas.

First Aid and Medical Treatment: First remove the poison latex from the exposed area by washing the area with soap and water. If there has been contact with the eyes, it is especially important to flush the eyes and surrounding tissues with water to remove the poison. There are no internal remedies; only local applications are used to soothe and reduce the inflammation. Typical topical remedies include calamine lotion, Vaseline, petrolatum, zinc oxide ointment, bismuth subnitrate ointment or topical steroids. The patient should be cautioned against scratching or rubbing the inflamed area to prevent secondary infection from developing as a result.

First aid and medical treatment for the ingestion of the fruit of this tree is the same as that for almost any other swallowed poison (see p. 5).

THE CASHEW TREE   (skin irritant sap)

The cashew nut tree, *Anacardium occidentale*, although a native of Brazil, has been introduced into many parts of tropical Central America. Like some other members of the Anacardiaceae family, it contains urishol-like resins that can cause contact dermatitis in susceptible individuals. The toxic principles can be found in all parts of the tree. The cashew ranges in size from a small shrub to a tree 40 feet tall with spreading branches. The cashew's leaves are leathery, yellow-green, oblong and four to eight inches long. The small pale yellow flowers develop into a kidney-shaped fruit about one inch long which is borne on a greatly enlarged fleshy red or yellow stalk, the so-called "cashew apple." In areas where the plants are numerous, some people enjoy eating the acid, tasty pulp of the cashew apple and others use it to make jams and drinks. The juice from the pulp and its rind can cause severe contact dermatitis in people who are sensitized to the urishol toxins of this family. The cashew nut itself has a smooth, olive-green, glassy outer shell that turns reddish as it matures. Between the inner and outer shell is an oil containing a urishol type oleoresin which is highly allergenic. In the production of edible cashews the nuts are roasted over flame to extract the oil and inactivate the poison, and to allow access to the delectable cashew nut familiar as a food. Many factory workers experience severe

23 Plants That Cause Dermatitis

Cashew Tree
Anacardium occidentale

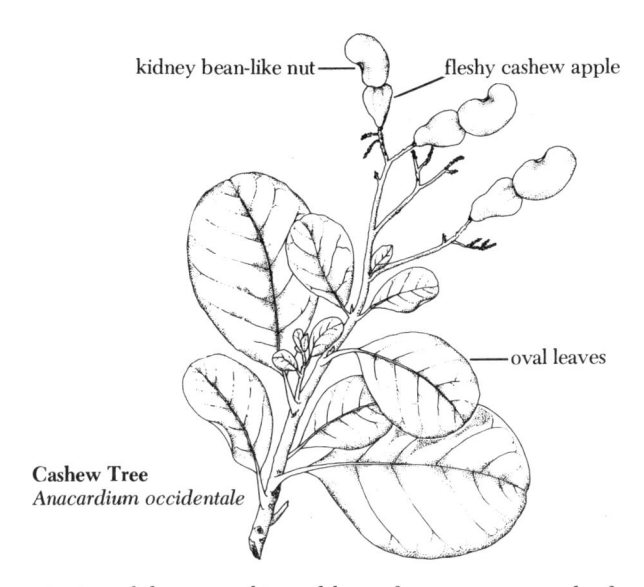

- kidney bean-like nut
- fleshy cashew apple
- oval leaves

irritation of the eyes, skin and lungs from exposure to the fumes from this process. The liquid, cardol oil, is used commercially to make plastics, electrical insulation, lubricants and insecticides. Eight million pounds of this oil are produced annually.

Toxins and Symptoms: All parts of the tree contain anacardic acid (90%) and cardol (10%) molecules similar to the urishols (oleoresins) of poison ivy. These tend to turn black when exposed to air and are usually inactivated by heat. The cashew nuts we get from commercial sources, even so-called "raw cashews," are usually totally safe due to heat treatment, but a recent (1983) mass poisoning resulted from a batch of improperly prepared or contaminated cashew nuts imported from Mozambique. Anytime you see cashew nuts with black spots, avoid them, since these spots are a good indicator of contamination. The symptoms of cashew contact dermatitis are exactly the same as those of poison ivy, and cross-sensitivity occurs. Thus, if you have a history of poison ivy allergic responses, you should avoid cashew nut trees. Be particularly careful of cashew wood fires, since droplets of the irritant in the smoke can cause inflammation of the eyes, nose, throat and skin. Eating cashew apples can cause severe dermatitis around the mouth in sensitive individuals.

First Aid and Medical Treatment: The same as for all Anacardiaceae; see p. 11.

24 *Poisonous Plants and Mushrooms of North America*

leaves with many leaflets

small flowers in round clusters

Polyscias
Polyscias fruticosa

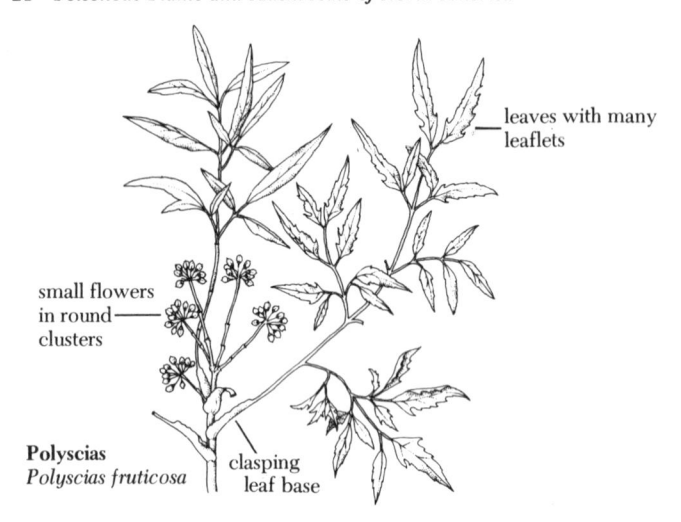

clasping leaf base

POLYSCIAS

A large, handsome, compact shrub, *Polyscias balfouriana*, of the ginseng family, is a common outdoor ornamental found in Florida. It grows as a 20-foot-tall shrub with many slender upright branches. The compound leaves have round, glossy green leaflets whose edges are variegated with white. There usually are three four-inch-wide toothed or saw-edged leaflets per leaf. During flowering, the plant produces large sprays of tiny white flowers. Several species of *Polyscias* are easily grown from cuttings and make excellent hedges which can be trimmed and shaped. Contact with cuttings is the usual cause of severe contact dermatitis and in southern Florida is a frequent source of dermatitis.

Toxins and Symptoms: The sap from this plant contains saponin and some unknown alkaloids. Repeated exposure to the sap apparently produces sensitization. When the sap comes in contact with the skin, or even when the skin is brushed against the leaves, skin irritation and rashes will ensue. There have even been cases of blistering and swelling.

First Aid and Medical Treatment: Wash off exposed areas with soapy water as soon as possible. See p. 11 for treatment.

BRAZILIAN PEPPER, FLORIDA HOLLY

In 1898 and several times thereafter, the U.S. Department of Agriculture introduced a handsome bushy tree *(Schinus terebin-*

thifolius) from Brazil for ornamental purposes. This pepper tree can reach 40 feet in height and canopy width. The leaves are compound, with a conspicuous red midrib, and five or more aromatic leaflets, each about three inches long. It has separate sexes, i.e., "male" and "female" trees, and the white, tiny flowers occur in clusters. The single-seeded fruits are rounded, pea-sized, red, aromatic juicy berries whose taste is peppery. The seeds (small, kidney-shaped and yellow) have been spread throughout warmer areas of Florida, and the Brazilian Pepper has become a real pest tree in some areas of the Everglades.

Toxins and Symptoms: The plant contains a highly volatile resin which is apparently abundant when the plant is in bloom. Any contact with the resin, such as handling incident to using the plant as a Christmas decoration, causes a rapidly developing itchy rash. Apparently the sap is an irritant rather than an allergen because the response is so rapid.

STINGING NETTLES

The stinging nettle, *Urtica dioica*, is a perennial Eurasian herb or weed found in almost all environments, particularly along moist roadsides and the edges of woodlands. It grows to a height of from two to six feet. The heart-shaped to oval paired leaves are up to six inches long, have prominent leaf veins and have saw-toothed or serrated margins. The leaves and stems are covered with sting-

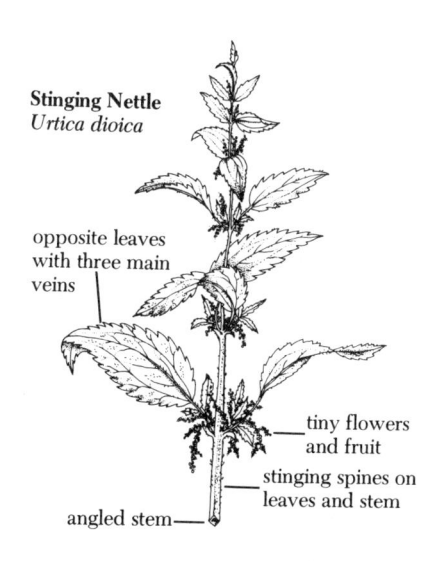

Stinging Nettle
Urtica dioica

opposite leaves with three main veins

tiny flowers and fruit

stinging spines on leaves and stem

angled stem

ing hairs, which are long needle-like structures with a broadened base. These hairs penetrate the skin and release their contents, causing severe local inflammation and irritation. The nettles' small, green flowers form long clusters arranged in loose, lacy bunches along a main axis and develop into inconspicuous small, dry achenes. (**Plate 1.**)

Distribution and Habitat: Stinging nettles are found throughout North America, usually in moist environments, and are particularly abundant along roadsides.

Toxins and Symptoms: The hollow hypodermic-like stinging hairs have a bladder-like base filled with some very potent irritants—acetylcholine, histamine and serotonin (5HT)—which have profound effects on blood vessels. When a hair penetrates the skin and is bent, the toxins are squeezed from the bladder, forcing them into the tube, which then breaks off in the skin. The symptoms manifest themselves almost immediately, and are characterized by an intense burning sensation, reddening of the skin, and itching. If exposure has been extensive and a large number of superficial blood vessels dilate, the victim may feel faint due to a drop in blood pressure, and in rare cases may go into shock. However, the main effect is inflammation of the skin, and is a self-limiting event that will correct itself in a few hours. Some susceptible individuals develop an immulogic response after repeated exposure to stinging nettle allergens. They may develop a classic contact dermatitis on subsequent exposure; erythema (reddening of the skin), itching, and small blister formation.

First Aid and Medical Treatment: First, avoid repeated exposure to the offending plant and protect exposed skin from other insults, particularly sunburn, which can worsen the symptoms. Apply cold water compresses to the irritated areas at half-hour intervals. Imbedded stinging hairs should be removed with sticky tape, and inflammation can be controlled with ½% to 1% hydrocortisone or other cortical steroid creams. Antihistamines such as Periactin or Temaril may help. If secondary infection occurs, treat as described on p. 11.

Related Species: Other species of nettles of the genera *Urtica* and *Laportea* as well as the spurge nettle, *Cnidoscolus stimulosus*, are equipped with stinging hairs. Stinging hairs are rather common in many plants of the tropics, and are characteristic of the Loasa Family, which is distributed in the American tropics.

PLANTS CONTAINING PHOTOSENSITIZERS THAT EXAGGERATE THE EFFECT OF THE SUN'S RAYS

A number of widely distributed common plants contain substances called psoralens (furocoumarins). These psoralens are readily released from bruised plants and on contact with the skin are rapidly absorbed. The psoralens are photosensitizers, that is, they are absorbed into the skin and trap the energy of the sun's rays at a variety of wavelengths, thereby producing a much greater response to solar radiation than would normally be expected. The symptoms are those of sunburn. Parts of the body not exposed to sun but exposed to the psoralen show no effect. People have different sensitivities to the sun's rays, and persons who are very sensitive and have contacted a photosensitizing plant can have very severe reactions in local areas such as over the brows, cheekbones, nose and chin. High humidity increases absorption of psoralens and increases their effects. Soapy washes after being in the field may remove some of the psoralens, and use of sunscreen creams may afford some protection from subsequent sun exposure. Some common photosensitizing plants are wild carrot (Queen Ann's lace, *Daucaus carota*), wild parsnip *(Pastinaca sativa)*, some meadow grasses, St. John's wort *(Hypericum* spp.), buttercup *(Ranunculus* sp.) and some citrus fruits.

PART II
Hallucinogenic Plants

THE word hallucination comes from the Latin *hallucinare*, which means to dream or wander in the mind. However, this is not a very precise definition. Today's psychiatrists describe hallucinations as sensory perceptions that occur in the absence of any real sensory stimulus and may involve any of the five senses. The underlying cause of such false perceptions is generally thought to be in the brain and to have been caused by some alteration of brain chemistry. Such changes in perception are common in some forms of mental illness, in people under stress, during sleep deprivation, during dreaming and in people using drugs that act on the central nervous system. A number of plants produce secondary chemical products that, if ingested by humans, will produce ecstasies, imagery, excitation and a variety of other altered mental states. Such plants are classified as hallucinogenic plants and their products as hallucinogens.

The imagery produced by hallucinogens has been described by a number of modern investigators. The majority of visual hallucinations induced by such compounds occur even if the eyes are closed and consist of symmetrical patterns, often saturated with intense colors, located at reading distance. The images are sometimes quite complex, and often include religious objects or animals. Reports of people experiencing hallucinogen intoxication include descriptions of vivid kaleidoscopic visions, religious experiences, feelings of unreality, distortions of time and space, delirium and in some cases panic reactions similar to those experienced in nightmares. The responses are generally predictable, but there is significant individual variation, and the experience may vary with a number of environmental variables including stress, crowding, fatigue, state of health, and exposure to alcohol or other mind-altering substances. It should be noted that plants often produce a variety of secondary chemical substances, and some of these are toxic rather than hallucinogenic. Furthermore, some plant hallucinogens, if taken in sufficient quantity, can cause a wide variety of severe side effects and in some cases death.

A BRIEF HISTORY OF HUMAN USE OF HALLUCINOGENIC PLANTS

Plants that produce hallucinations that divorce the mind from reality have a long history of human use, probably preceding recorded history. Their effects were given magical or mystical and religious meaning, and their usage was thought to promote growth of the spirit, the experience of a new reality and enhancement of all perceptions, including the experience of food, drink, and sex. Only since the 1950s have the plants and their products been utilized as purely recreational mood alternators. There are two general groups of hallucinogenic plants, those whose secondary chemical products contain nitrogen and those whose products contain no nitrogen.

Among the non-nitrogen-containing hallucinogens is one of the oldest and most commonly used mind-altering plant, marijuana, *Cannabis sativa* (see p. 32 for details). This plant was called by the ancient Chinese the "Liberator of Sin" and the "Delight Giver." The Hindus called it "Soother of Grief" and "Heavenly Guide." In the Middle East and Africa, parts of the plant or extracts were called "hashish" from whose origins come the words assassin, "bhang," and "ganja." In the United States, where the dried plant is smoked in pipes or cigarettes, the terms reefer, pot, grass, and many others are commonly used.

The other non-nitrogenous hallucinogen comes from the nutmeg tree, *Myristica fragrans*. While this plant is commercially grown in the West Indies, the main source of the hallucinogen is commercially available dried seeds.

The majority of hallucinogenic plants and fungi have nitrogen in their psychoactive chemical substances. These include the peyote cactus *Lophophora williamsii* (p. 30), common in the southwest United States and Mexico. Some South American snuffs which are not included, the hallucinogenic mushrooms of the genera *Psilocybe*, *Conocybe* and *Stropharia*, are discussed in detail in the section on mushrooms. Morning glories *(Ipomoea tricolor)* and their kin, a number of members of the genus *Datura* (jimsonweed), the opium poppy, and a few others will be reviewed in the following pages.

There are also a number of fungi, the magic mushrooms, that contain potent hallucinogens and have been used by humans for centuries. These we have chosen to list in the mushroom section beginning on p. 131.

Our knowledge about the relative toxicity of these compounds and both the short- and long-term effects of their usage is sketchy. The controversy regarding the legalization and use of a number of these compounds is a matter of ongoing legal action. In this book we have by intent made no judgments about their use but have tried to provide the most up-to-date descriptions of what is currently known about toxicity, symptomatology and treatment for intoxication caused by these plants. Cultivation and possession of some of these plants is outlawed in certain states.

PEYOTE CACTUS

Peyote cactus, *Lophophora williamsii*, is a plant that contains a hallucinogenic substance that has been used in the New World since pre-Columbian times. As early as the 16th century, the conquistadors recorded that the Aztecs and other Mexican tribes used this plant both as a medicine and as a hallucinogen associated with mystical religious experiences. Indeed, the use of peyote in religious rituals spread from the Mexican Indians to tribes in the United States and southern Canada. Even today the Native American Church, with close to a quarter of a million members, incorporates the use of peyote into their religious ceremonies. Peyote is referred to by a number of colloquial names: peyote, peyotillo, divine herb, mescal button, devil's root and medicine of God.

Peyote is usually an inconspicuous, blue-green, spineless cactus whose surface appearance is as a small (two inches in diameter) flattened button with a depressed center from which emerge four to 14 radial furrows. Each section between these furrows has a cluster of yellowish-white hairs. The root, the largest part of the cactus, is turnip-shaped. While this description is adequate for small wild plants, in areas where harvesting of peyote buttons is common, the multiple-headed clumps may measure six feet across. Peyote flower buds emerge from the middle of each head and develop into large pink to white flowers with many petals. The pink to red cylindrical fruits (an inch-and-a-half long) have flat, warty black seeds. As the fruits mature, they become dark

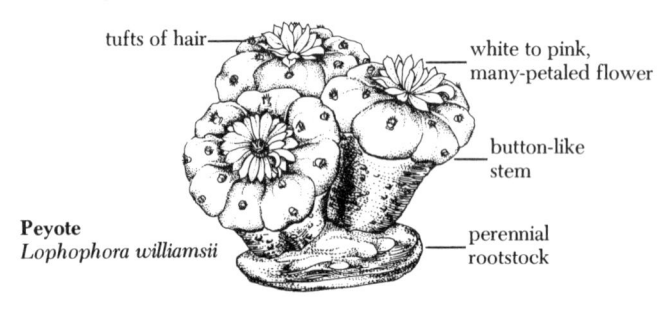

Peyote
Lophophora williamsii

tufts of hair
white to pink, many-petaled flower
button-like stem
perennial rootstock

31 Hallucinogenic Plants

brown; the walls become thin and fragile and rupture, releasing the small black seeds for dispersal.

Distribution and Habitat: The plant is widespread through the Chihuahua desert, the Rio Grande region of Texas, in West Texas in the Rio Grande valley near Safter and Laredo, southward into Mexico in the basin between the Sierra Madre Occidental and the Sierra Madre Oriental. It grows under diverse conditions in this desert, ranging in elevation from 150 to 5,600 feet. It is often found under mesquite, creosote bush and *Agave*, but can also thrive in open areas devoid of shade or protection.

Toxins and Symptoms: There are currently more than 55 known secondary substances found in peyote but the most important of these alkaloids is mescaline, a trimethoxy indole, a compound whose structure resembles two important neurotransmitters found in the human nervous system, dopamine and norepinephrine. This potent, biologically active substance (mescaline) makes up about 1 percent of the total dry weight of the plant. Human ingestion is usually by eating dried tops or mescal buttons, or the buttons may be powdered and made into an herbal tea.

An hour or so after ingestion of peyote, one experiences a spectrum of rather unpleasant symptoms: nausea, dizziness, sweating, vomiting, palpitations of the heart, and a variety of pains in the chest, head and neck, often accompanied by hot and cold flushing, shortness of breath, a minty taste in the mouth, abdominal cramps, a sense of urinary urgency, pupillary dilation, tremors, and restlessness. Some individuals experience severe anxiety or fear, panic reactions which on rare occasions cause the user to become aggressive and even assaultive. These early responses to mescaline may last two to four hours and gradually develop into a psychic phase whose symptoms are dominated by a sense of well being, elation to the point of ecstatic euphoria, a sense of physical power and vivid fantasies that range from realistic daydreams to full-blown hallucinations. All sensory inputs are distorted, particularly visual perceptions. These sensory phenomena and thoughts follow in rapid succession, and while the user may be unable to express himself clearly he usually remains awake and lucid, and does have recall of the experiences. This "high" may last for several hours, and then the user usually falls asleep. The whole experience, from the acute unpleasant symptoms through the high stage, may last eight to ten hours.

During the high phase, the pupils are dilated, there is some loss of motor coordination, the rate of breathing increases and sweating, blushing and salivating usually take place. The entire spectrum of responses is dependent on a number of factors, including amount ingested relative to the weight of the user, the personality of the subject and his mood and physical state at time of ingestion. Usual mescaline doses for a "high" are five milligrams per kilogram of body weight, or about one-third of a gram for a 160-pound human. It is not particularly toxic, and humans have ingested as much as eight grams and survived, although they got very sick during the first phase of the reaction.

First Aid and Medical Treatment: Since most mescaline users vomit during the first phase of the reaction, it is not necessary to induce vomiting. Additional pharmacological intervention is probably not a good idea, and therapy should be in the form of emotional support. Provide the victim with parenteral fluids if vomiting has been extensive. Treat severe agitation with Benzidiazipines. Usually Dr. Time brings about a full recovery within 24 hours or so.

CANNABIS, MARIJUANA

One of the oldest known hallucinogenic or mind-altering plants is *Cannabis sativa*, also known as hashish, marijuana, bhang, kif, grass, pot, or hemp. The list of names for this plant is as long as the controversy that surrounds its use, a use that was recorded in China about 10,000 years before Christ. As early as 2000 B.C., ancient Indians used a concoction called "bhang" which was

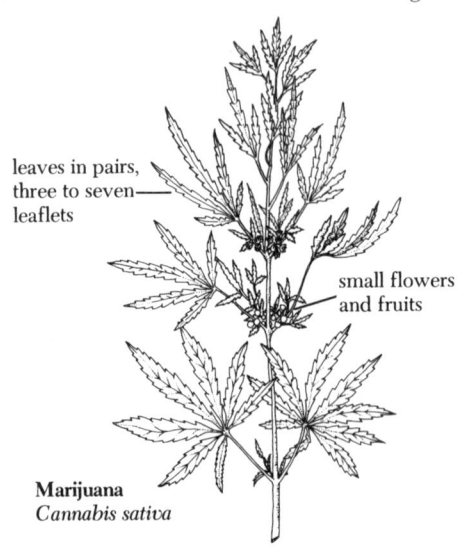

leaves in pairs, three to seven leaflets

small flowers and fruits

Marijuana
Cannabis sativa

thought to release people from anxiety. The plant may be one referred to in the Old Testament (Exodus 30:23). The word "assassin" originated with a group of religious terrorists in Persia who were known hashish users (hashshashin), although the Arabs, who have used *Cannabis* for centuries, don't view it as a violence-producing substance. In 1484 the Pope listed *Cannabis* as one of the drugs witches used to summon the Prince of Darkness. However, it was not until the early 19th century that *Cannabis* became popular among some writers and artists in Europe. Hemp was brought to the Americas with the early settlers, mainly as a source of fiber for ropes, but by the mid-1800s the mind-altering properties of pot had brought this plant into wide recreational use. In 1883, according to an article in *Harper's* magazine, there were 600 hashish houses in New York City. Subsequently, the use of hash declined, and laws banning its sale and use began to appear in some states. National legislation against marijuana began in the 1930s, mainly as a ploy to get more funding for the Federal Bureau of Narcotics.

Cannabis, a member of the family Cannabinaceae, probably originated in Central Asia, but is now grown everywhere. *C. sativa* is an annual weed that may grow to 15 feet in height. There are both male and female plants, with the male being somewhat taller than the female. Both are easily identified by their large, long stalked, palmately compound leaves. Each leaf has five to seven slender lance-shaped leaflets whose margins are toothed. The small green flowers are borne among the leaves, with the male flowers in open clusters roughy two-thirds of the length of the leaves, and the female flowers being borne in short leafy groups alongside the stem. *C. sativa* stem fiber is extremely strong, and has been used to make rope and canvas; indeed, the word canvas derives its origin from the name of the genus *Cannabis*, and Levi Strauss' original jeans were made of hemp sailcloth. However, in today's world, the main commercial use of *C. sativa* is as a source of mind-altering compounds which are most concentrated as a resin in the flowering tops of cultivated female plants. (**Plate 2.**)

Distribution and Habitat: Widely distributed *Cannabis* can and does grow on both wastelands and fertile lands as a weed. Various races and hybrids rich in resin are cultivated in isolated plots, gardens, and greenhouses. Cultivation of marijuana is illegal and can lead to prosecution.

Toxins and Symptoms: There are a large number of biologically active compounds found in *Cannabis*. Their general name is cannabinoids, and these include, among others, cannabinol, cannabidolic acid and tetrahydrocannabinol (THC), the last being the main mind-altering compound. There are also alkaloids, phenols and other compounds present in varying amounts, but their effects are not well understood. There are different races of C. *sativa* which contain differing concentrations of THC. Those races with the highest resin concentration are the most valued by users. The main route of administration which gets the highest concentration of THC rapidly into the blood is via the smoke of burning pot, usually as a cigarette or joint. An experienced smoker can absorb 50 percent to 75 percent of the available THC in a joint. As little as five milligrams can produce measurable effects, which may begin in five to 15 minutes. Peak effects are reached in about 60 minutes and may persist for several hours. There is some controversy over the biological half-life of THC (the time for half the absorbed drug to be eliminated), but it is known that THC stays in the body for a relatively long time (24 to 56 hours to eliminate half and another 24 to 56 hours to get rid of half of the remaining half). The major route of elimination is the bile and feces. Marijuana is not physiologically addictive, but some users do develop a psychological dependence on the effects produced. While popular myth suggests that a user does not develop tolerance to THC, most clinical investigators feel that tolerance does develop rather quickly. This means that the user must use progressively greater amounts of pot to achieve a "high." Generally, marijuana is nontoxic in that it can produce its mind-altering effects in concentrations 4,000 times less than the amount that would be lethal (about 10 grams per kilogram of body weight).

Surveys have reported widespread use of marijuana, although since 1978 there has been a consistent decline in its use. However, 15 percent of junior high school students and 40 to 50 percent of high school and college students report trying it at least once. Our knowledge of marijuana effects, particularly long-term, is rather limited, but a recent (1982), very careful study by the National Academy of Science concluded that THC "is not innocuous" and its use "justifies serious national concern."

Like cigarettes, joints cause heavy smokers to develop chronic inflammation of the upper respiratory tract, bronchitis, pharyngitis, sinusitis, asthma and possibly even pre-cancerous changes in the lungs. Pot smoke can irritate the eyes, causing conjunctivitis. THC is known to lower pressure within the eyeball and has been used to treat people suffering from glaucoma, the second leading cause of adult blindness. Long-term effects of heavy marijuana smoking on the heart and blood vessels have not been reported, but we know smoking tobacco heavily does increase the risk of heart disease in both men and women. Women smokers have 20 times greater risk than non-smoking females. THC is

35 Hallucinogenic Plants

known to suppress male hormone production and decrease the size and weight of the testes and prostate. In the 1930s, males of college-age had sperm counts that averaged 90,000,000 sperm/cc, while male students in the early 1980s had average counts of only 60,000,000/cc. This may have been due to the effects of tight jeans or pot smoking, and fortunately these effects are usually reversible. One study on young males reported breast development in heavy smokers, but there may have been a selective sampling error in this study.

In pregnant females, THC crosses the placental barrier and reaches the fetus. It also may be secreted in mother's milk and reach the newborn. While there are no reports of birth defects in human pregnant pot smokers, animal studies do report an increase in such defects; hence caution during pregnancy and lactation is recommended. THC also produces a temporary and reversible depression of the body's immune defense system, which could possibly increase the risk of infection and cancer.

The major effects of *Cannabis* compounds are on the brain and behavior, but these effects are controversial and not well understood. Its well-described effects on sensory and perceptual functions are the principal reasons for its use. Recent research findings have reported that THC also impairs motor coordination, interferes with short-term memory and impairs the learning process in most subjects, possibly by decreasing the drive or motivation to master a task. Controlled studies on human subjects have included a number of recurrent symptoms such as loss of ambition, apathy, difficulties in concentration, decrease in performance and loss of efficiency. There are also reports on subjective effects: elevated feelings and enhanced pleasure reactions to everything from food to sexual experiences. THC also lowers aggressiveness, decreases feelings of anxiety and frequently produces a sense of euphoria, a "high" often accompanied by "the giggles." Panic reactions have been reported by some users, but this type of response is relatively rare. Overall, the use of *Cannabis* does not seem to produce acute damage to the body, but we still don't understand the mechanism of action of THC, and although there is no solid evidence of long-term hazards associated with its use, this does not mean that there are no such effects.

MORNING GLORY (seeds) rarely ☠

The early Spanish historians who chronicled the conquistadors' Mexican experiences wrote of the religious use of the lentil-like seeds of a flowering vine which we know today as the morning glory. Morning glories (*Ipomoea tricolor*) and the sacred Aztec hallucinogenic plant, ololiuqui (*Turbina corymbosa*) are both

Morning glory
Ipomoea tricolor

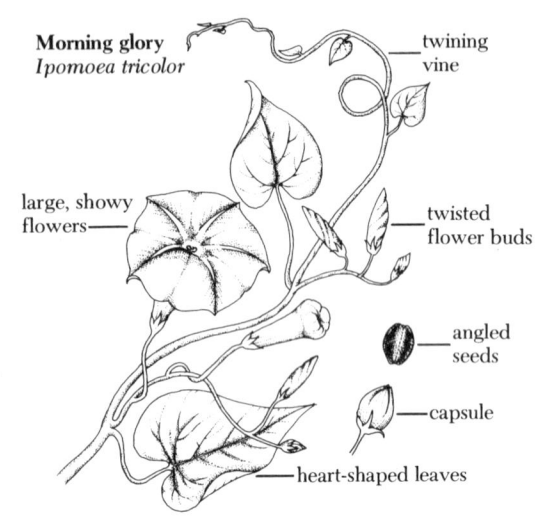

- twining vine
- large, showy flowers
- twisted flower buds
- angled seeds
- capsule
- heart-shaped leaves

classified in the Convolvulaceae. Four hundred years ago, the conquistadors banned the use of these "diabolic seeds" because the users experience ". . . a thousand visions and satanic hallucinations." During the drug explosion of the 1950s and '60s, once it became known that morning glory seeds contained enough LSD-like compounds to produce hallucinations, the local garden shop's shelves were emptied by enthusiastic experimental pharmacologists, but the seeds contained other alkaloids as well, and several serious cases of physical damage and death followed overdoses. Some studies with seeds produced no effects, probably because the tough outer seed coat has to be broken or pulverized to release the active components. While *Ipomoea tricolor* seeds are known to contain high amounts of active alkaloids, recent research has suggested that several other members of the genus *Ipomoea* also have toxic contents.

Morning glories, both wild and cultivated ornamentals, are common twining vines with tendrils. The stalked alternate leaves are broadly ovate or heart-shaped and range in size from three to five or more inches in length. Their large, funnel-shaped flowers open in the morning; hence their name, and close by the afternoon. The flowers are blue, white or purple-pink and may range in size from two to five inches in diameter. The vines, from ten to 20 feet long, bear large numbers of round capsular seed pods, each containing many seeds. Among known potent varieties are "Pearl Gates" and "Heavenly Blue," but even wild morning glories such as *Convolvulus arvensis* contain toxins in their seeds. **(Plate 3.)**

Toxins: The seeds contain a number of indole alkaloids, such as d-lysergic acid, ergoline and other compounds very similar in structure and action to the notorious hallucinogen LSD. The plants may also contain toxic concentrations of nitrates.

Symptoms: Making a brew of 50 or more pulverized seeds of various morning glories can cause nausea, vomiting, blurred vision, loss of motor coordination and a variety of psychotropic effects such as hallucinations similar to those described for the magic mushrooms (see p. 150). There is considerable variation in the alkaloid content of different species; hence symptoms can be highly variable. Potent brews of more toxic species have caused convulsions, coma, and death.

First Aid and Medical Treatment: Be prepared to treat for hypertension and, if severe vomiting has occurrred, to provide fluid replacement therapy. If the patient is exhibiting panic reactions give Benzodiazipines.

THORN APPLE, JIMSONWEED, (all parts)
JAMESTOWN WEED dangerous

The thorn apple, *Datura stramonium*, an imported native of Asia and a member of the nefarious nightshade family (Solanaceae), is common and widespread in the United States, particularly in the South. The genus *Datura* has a long history of use as a hallucinogen in Asia and the Americas, where it was employed for pleasure, for sacred rites and as a narcotic medicine. Its preparation for sacred rites varied with different tribes, being smoked, powdered as a snuff or made into brews. One South American Indian tribe that buried wives alive with their deceased husbands used *Datura* extract to induce a stupor in these unfortunate women prior to burial.

Jimsonweed derives its name from an episode of mass poisoning of soldiers at Jamestown, Virginia, in 1676; hence the name Jamestown weed, which was corrupted to jimsonweed. In recent years children have died from eating its seeds and even from sucking nectar from the trumpet-shaped flowers.

The most common local representative of *Datura*, the thorn apple, is a stout, erect annual weed about five feet tall. Many branches emerge from a thick stem and form a rather showy spread of green or purple fetid-smelling leaves. The five- to six-inch-long leaves are alternate, simple, smooth, broadly ovate and unevenly toothed. The plant produces trumpet-shaped, pale violet to white, four-inch-long, five lobed, upright flowers, which eventually give rise to oval, two-inch-long, thorn-covered hard fruits with many seeds; hence the name thorn apple. (**Plate 4.**)

tubular, five-pointed flower

large leaves with irregular lobes

spiny capsule

Thorn Apple, Jimsonweed
Datura stramonium

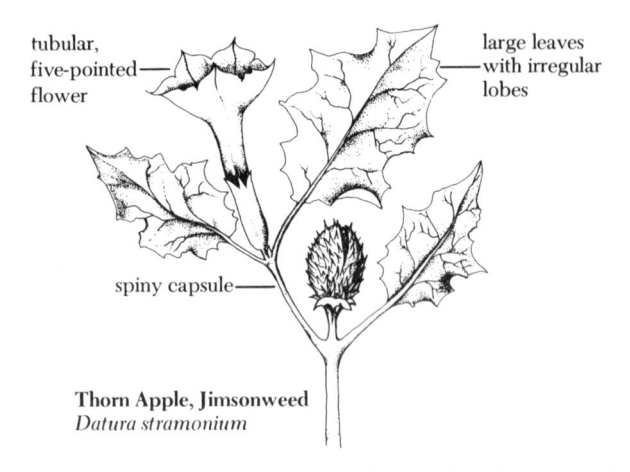

Distribution and Habitat: Thorn apples are found as an invader of cultivated fields and wastelands and sand dunes across much of the United States, particularly in the South and Southeast.

Toxins and Symptoms: Members of this malignant genus produce a number of alkaloid toxins—tropanes—which include atropine, a drug which acts to block acetylcholine receptors; hyoscyamine and hyoscine (scopolamine), with the latter being best known as a hallucinogen. A dilemma involving the use of jimsonweed seeds, roots or leaves (all parts are poisonous) for recreational use is that if you take a quantity of the hallucinogen that will do no perceptible harm, you may not experience any central nervous system effects, but if you increase the dose of atropine necessary to produce the effect you may reach toxic levels. Consequently, there is the risk of serious poisoning and death with each hallucinogenic experience. The signs and symptoms depend on the relative concentration of tropane alkaloids in the plant, the amount of plant material ingested and the mode of ingestion. Brews or other liquid concoctions made from plant parts are the most potent and fast acting.

The symptoms often present themselves in an orderly sequence. To quote Dr. J.M. Arena, "Hot as a hare, blind as a bat, dry as a bone, red as a beet and mad as a wet hen" summarizes them quite picturesquely. Initially, the nose and mouth are dry, the patient is thirsty, pupils dilate and looking at light is unpleasant, salivation stops and speech becomes difficult. Vomiting is unfortunately infrequent; the skin is dry, hot and flushed; sweating is suppressed; body temperature rises to 104° or more and vision becomes blurry. There is often a loss of coordination, frequently accompanied by irrational behavior, incoherence, and hallucinations. In severe cases of intoxication, seizures, coma, and death have occurred.

First Aid and Medical Treatment: The major problem with jimsonweed intoxication is that most of the time the victims of ingestion don't vomit. If the patient is alert, induce vomiting. If the patient is drowsy, get him to a hospital for gastric lavage, activated charcoal and cathartics. (Note to physician: Because of the dryness of the throat, it may be necessary to lubricate the stomach tube very well.) Administer 2 mg of physostigmine I.V. to counter the anticholinergic action in the central nervous system and repeat in cases of severe intoxication, since physostigmine is rapidly degraded by the body. *DO NOT* give phenothiazines, since they can exacerbate the effects of the atropine component of the plant and cause cardiovascular collapse. Barbiturates or Benzodiazipines may be called for if the patient is convulsing, and it may be necessary to use external cooling to lower the body temperature. Monitor blood pressure for hypertension and monitor EKG for possible ventricular arrhythmias.

ANGEL'S TRUMPET, DEVIL'S TRUMPET (all parts) dangerous ☠

There are a number of ornamental species of the genus *Datura* that bear the common names "angel's trumpet" or "devil's trumpet." All are potentially deadly toxic plants that have been used for recreational purposes. In Florida there have been a number of hospitalized cases of intoxication and two deaths of teenagers who concocted a brew from the flowers of one of the angel trumpets, *D. suaveolens*.

Datura suaveolens, a handsome tree-like shrub or large herb popularly used in the Southeastern states as an ornamental evergreen, stands up to ten or 15 feet tall. It has long (up to 12 inches), oblong, dull green leaves and very showy large (eight to 12 inches long), horn-shaped, white, downward-hanging flowers that emit a sweet, musky scent. Usually, the plant does not form fruit, but if it does the black seeds inside the spindle-shaped pod are highly toxic. All parts of the plant are poisonous, but in practice usually only the flowers are used to make hallucinogenic teas. One myth is that removing the pistils from the flowers renders the brew safe. This is not true.

Distribution and Habitat: This native of Brazil thrives outdoors in frost-free regions of the United States, mainly California and

Florida, where it has been responsible for large number of cases of intoxication and a few deaths. The other species of *Datura* called angel's trumpet are:

1. *Datura metel*, the hairy thorn apple, or devil's trumpet, and the related *D. meteloides* are tall annual herbs with large, foul-smelling leaves that produce large (up to eight inches long), erect, trumpet-shaped flowers that are purplish-white on the outside and yellow or white on the inside. The flower stems may be purple, and the pendulous, round seed pod is covered with short spines. The seeds are sweet tasting, flat and brown.
2. *Datura candida* (often erroneously named *D. arborea*) is a small tree (up to 15 feet tall) similar in appearance to *D. metel* except that its leaves and flowers are longer and the flowers are pendant-like. The trumpet-shaped, white flowers are very fragrant, and the smooth seed pod is elongate, egg-shaped and two-and-a-half inches long. The plant is common in Florida and Hawaii.
3. A hardy angel's trumpet is *D. sanguinea*, a short tree-like shrub (up to twelve feet tall) with seven-inch-long leaves, tube-like, ribbed, red or orange pendant flowers and a conical, smooth seed pod. This plant is common in California and may be found in greenhouses in colder climes.

Other species of *Datura* should also be considered poisonous. The names of these species are confused at present, with certain species sometimes included in the genus *Brugmansia*.
BEWARE OF ALL TRUMPET-SHAPED FLOWERS OF *DATURA*.

Signs and Symptoms: See jimsonweed, p. 38.

First Aid and Medical Treatment: See jimsonweed, p. 39.

THE MATRIMONY VINE (all parts) rarely ☠

The matrimony vine, *Lycium halimifolium*, sometimes called "the box thorn," can grow either as a spreading shrub or a woody vine. It has a multibranched, thorny gray stem and alternate, lance-shaped, two-inch-long leaves that are grayish-green on their undersides. The solitary, usually purplish trumpet-shaped flowers with five distinct lobes later develop into jelly-bean-sized reddish-orange berries with many seeds. A native of Eurasia, the matrimony vine was introduced into this country as an ornamental, but it escaped and now grows wild. All parts of the plant are toxic, and cattle and sheep have been poisoned by eating the leaves and young shoots. The attractive flowers and berries may attract young children and cause intoxication.

Toxins and Symptoms: This member of the nightshade family contains tropane alkaloids similar to hyoscyamine and has an atropine-like action. Symptoms would be similar to those produced by jimsonweed (see p. 38).

First Aid and Medical Treatment: See p. 39.

Opium Poppy

Of all the members of the poppy family, the opium poppy, *Papaver somniferum*, is the most famous. Although all poppies contain similar alkaloids, the opium poppy has been cultivated because it produces the most raw opium per plant. Opium comes from the Greek *opion*, meaning juice, and the name is applied to the milky latex that oozes from cut, unripe seed pods. This gummy juice was used medicinally as early as the third century before Christ as a cure for diarrhea. In the 17th century, opium was introduced into China by the British and was ultimately the cause of the bloody Opium Wars that occurred two hundred years later.

The opium poppy is an erect, two- to four-foot-tall annual herb with alternate, up to ten-inch-long, heart-shaped, toothed, hairy leaves that clasp the stem. The handsome, showy, four or more petaled, solitary flowers, usually red or orange, can vary in color depending on the variety. The flowers are large, with dark centers

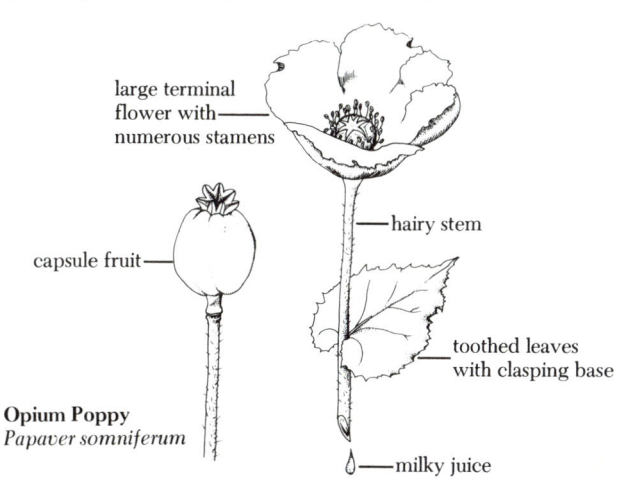

large terminal flower with numerous stamens

capsule fruit

hairy stem

toothed leaves with clasping base

Opium Poppy
Papaver somniferum

milky juice

and two or three sepals that fall as the petaled flowers open. After flowering, the plant forms a walnut-sized, smooth capsule filled with small black seeds and covered with a wheel-like cap. (**Plate 5.**)

Distribution and Habitat: Opium poppies and related species are widespread across the United States, southern Canada and Central America. This Eurasian import was once commonly cultivated, but now cultivation is a U.S. federal crime. Poppies grown for the illicit drug trade (morphine and heroin) are found mainly in the Middle East and Asia.

Toxins and Symptoms: There are a number of potent alkaloids present in opium gum, the most important of which is morphine, which constitutes about 10 percent of the alkaloid content. Morphine is named after the Greek god of dreams, Morpheus. These opioids are closely related to two recently discovered chemicals, the endorphins and encephalins, which are found in the brain and other parts of the human neuroendocrine system. These chemicals react with three different types of receptors to produce a spectrum of effects which are mimicked by the poppies' alkaloids. These responses include: suppression of pain (analgesia), respiratory depression, mood elevation (euphoria), pinpoint pupils (miosis), strong sedation, inhibition of bowel motility, inhibition of smooth muscle of the female reproductive system, inhibition of the cough center, stimulation of the vomiting center and production of a dreamlike state that imitates hallucinations. Medically, opioids have been used to treat pain, for sedation, to inhibit coughing and prevent diarrhea, although an aftermath of their use is constipation. Their recreational use has been to produce euphoria, dreams, and generally as a chemical escape from the stresses of life; however, opioids are strongly addictive, and withdrawal from use of opioids is a horrendous experience. In large amounts, opioids are toxic, causing death by respiratory depression. All parts of young poppy plants contain opioids and should be considered poisonous to eat, although to reach toxic quantities considerable plant material must be ingested. The seeds are the least toxic, although many of us remember getting high as children on the poppy-seed-filled triangular pastries our grandmothers used to make.

First Aid and Medical Treatment: If patient is alert induce vomiting and if drowsy use gastric lavage, activated charcoal and cathartics. Monitor respiration and be prepared to treat respiratory depression as needed. Nalaxone is an effective antidote, 0.03mg/kg body weight for children and 1.2 mg for adolescents and adults. This treatment may be repeated once.

43 Hallucinogenic Plants

ATROPA BELLADONNA   (all parts) rarely ☠

Atropa belladonna, another villain of the nightshade family, has been used as a narcotic for centuries and was brought to this country from Eurasia for its medicinal value. *A. belladonna* grows wild, is cultivated in herb gardens and is also used as an ornamental. Its name is derived from the mythological *Atropos*, one of the three Fates, who was responsible for severing the thread of life. It is a perennial herb that sprouts a branched, upright stem three to six feet tall. *A. belladonna* has ovate, glossy, alternate or opposite leaves without marginal teeth and solitary drooping five-lobed, bell-shaped purplish flowers with large, persistent sepals. It has cherry-sized, round, black, many-seeded berries that exude a violet juice when squashed. The name belladonna, which in Italian means "beautiful woman," comes from the use of its juice as a cosmetic to enlarge the pupils of the eyes. Dilated pupils usually signal approach. It has been speculated that the allure of the famous painting, the Mona Lisa, is in part due to atropine-induced dilated pupils. The mysterious hint of a smile, however, is purely her own. All parts of the plant are toxic, with the greatest concentration of toxin in the seeds, roots and leaves.

Toxins and Symptoms: This plant contains hyoscyamine and atropine; hence its actions are very similar to those described for *Datura* (see p. 38), although its major action is atropine-like.

First Aid and Medical Treatment: See p. 39 (*Datura*) for treatment.

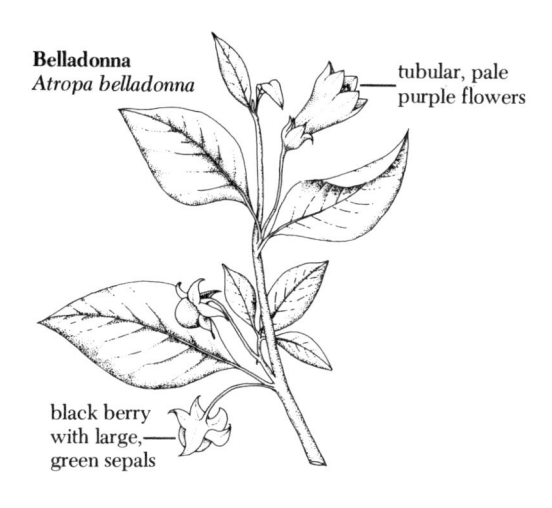

Belladonna
Atropa belladonna

tubular, pale purple flowers

black berry with large, green sepals

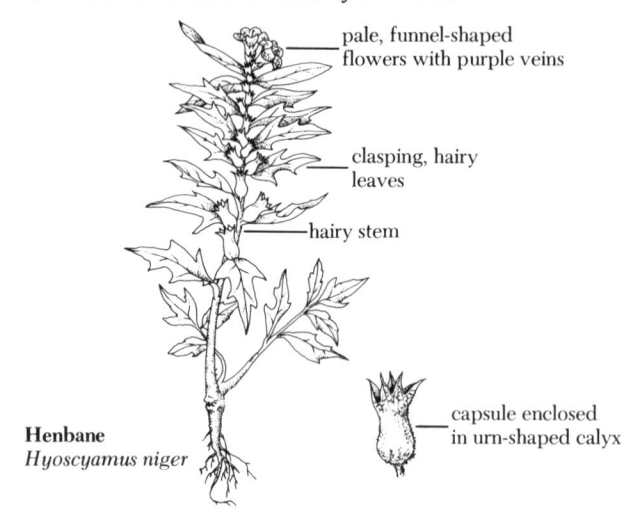

Henbane
Hyoscyamus niger

- pale, funnel-shaped flowers with purple veins
- clasping, hairy leaves
- hairy stem
- capsule enclosed in urn-shaped calyx

BLACK HENBANE (juices, seeds) dangerous ☠

Black Henbane, *Hyoscyamus niger*, also called fetid nightshade, insane root and poison tobacco, is another foul-smelling, toxic member of the nightshade (Solanaceae) family that has a long history of medicinal use dating back to the ancient Egyptians. The scientific name derives from Greek, meaning "hog bean," associated with its use in poisoning wild pigs. This annual or biennial, 30-inch-tall weed has a central, erect, hairy stem with many hairy branches. Its leaves are five to seven inches long, alternate, clasping, somewhat oblong in shape and unevenly toothed or lobed. The midribs of the leaves bear long hairs. The one- to two-inch-wide yellowish flowers are trumpet-shaped with prominent purple veins, and are produced among the leaves or in a leafy spike. The fruit is a round, half-inch-long capsule containing many seeds and enclosed in the large, long-persisting sepals. The plant contains the same toxic ingredients as the thorn apple (see p. 38), and produces similar effects. It has been responsible for poisonings, particularly in children who have eaten the seed pods or seeds.

Distribution and Habitat: Henbane is grown in herb gardens and has escaped to disturbed ground in dry, sandy soils across the northern United States and southern Canada, but it is particularly common in the Northwest.

Toxins and Symptoms: Similar to *Datura;* see p. 38.

First Aid and Medical Treatment: See p. 39 for treatment of anticholinergic poisons.

PART III
Toxic Plants of the Home and Garden

BECAUSE of the enormous impact of toxic wild plants on horses, sheep, cattle and other browsing mammals, quite a bit is known about their toxicity and the nature of their toxins. House plants and garden ornamentals have not been as carefully studied, yet their use has increased tremendously in the past two decades. With this increase there has been a parallel increase in the number of young children experiencing toxic distress, life-threatening emergencies and even deaths following ingestion of some of these beautiful but potentially deadly plants. The National Clearinghouse of Poison Control Centers reported a 20 percent increase of toxic plant ingestions from 1975 to 1976. There were a total of 7,710 cases of ingestion, of which 186 required hospitalization, and of these, three died. Unfortunately, there is a very uneven quality to the existing information, and most experts agree that these numbers are actually only the tip of the iceberg.

Should toxic or potentially toxic house plants bear hazardous warning labels? Certainly tobacco products, insecticides and a variety of household chemicals bear warning labels, and there have been attempts to put hazard labels of potentially dangerous plants used in interstate commerce. These attempts have been vigorously resisted by commercial horticultural associations and nurserymen for obvious reasons. However, they, the nurserymen, can make a case since the scientific literature on toxic house plants is rather imperfect. A case in point is the unwarranted claims that the popular Christmas decorative plant, the poinsettia, is poisonous (see p. 97). Naturally, the poinsettia growers, whose livelihood is threatened, reacted with great vigor. Nevertheless, there are some ornamentals that are known to be toxic, and the problem is drawing the line on how toxic a plant has to be to earn the label "hazardous." The plant growers put forth a simple defense, which is: labels aren't needed, and people should be smart enough to not eat anything they don't recognize as wholesome. However, children, particularly those under five, are the most frequent

46   Poisonous Plants and Mushrooms of North America

victims, and parents should be made aware of the potential hazard. Until the federal or state governments pass a law, the majority of the public will not be protected. One important function of this book is to provide concerned people with the most up-to-date information on poisonous house plants. To this end, we have made an extensive review of the literature and tried to synthesize this information into a form that is highly readable and understandable to both the layman and practicing clinician. For those wishing to dig deeper into the subject, we have included a collection of selected references at the back of the book.

LILY FAMILY   (bulbs) rarely ☠

This lovely lily family contains some of our most exquisite ornamental garden flowers and house plants, such as the lily-of-the-valley *(Convallaria)*, the climbing lily *(Gloriosa)*, asparagus, hyacinth, the autumn crocus or meadow saffron *(Colchicum)* and the star of Bethlehem *(Ornithogalum)*. Adorning our woodlands are large numbers of wild flowers of the lily family, among which are the false hellebore *(Veratrum viride)*, the black snakeroot or death camas *(Zigadenus)*, fly poison *(Amianthium muscaetoxicum)*, dogtooth violet *(Erythronium)*, bear grass *(Nolina)*, Solomon's seal *(Polygonatum)*, wild lily-of-the-valley *(Maianthemum canadense)* and Indian cucumber root *(Medeola virginiana)*. While these plants satisfy our visual appetites, only sickness and possibly death may result if we use any of these plants to satisfy our gastronomic curiosity—these plants contain a number of poisons, the most important of which are alkaloids and cardiac glycosides. After a child has eaten the plant, these poisons can have a very direct effect on his nerves and muscles, leading to gastrointestinal upset, problems with respiration and the circulatory system and, in severe cases, coma and even death. Fortunately, members of this family are readily recognized and easily distinguished from most edible plants. Species in the lily family have leaves with parallel veins; these leaves may vary from long and grass-like to short and oval. The leaves may be produced in a tuft at ground level or on a leafy stem. The root system is often an onion-like bulb, but may also occur as a creeping horizontal rootstock. The flowers have six petal-like parts, produced in two whorls and usually six stamens. The fruit can be either a several-seeded berry, as in most of our wild flowers, or a dry capsule.

Onions and asparagus, also members of the lily family, have minor amounts of potentially poisonous compounds. Eaten in normal quantities, these plants do not pose any health risk; however, accidental poisoning may come through confusing poisonous

47 Toxic Plants of the Home and Garden

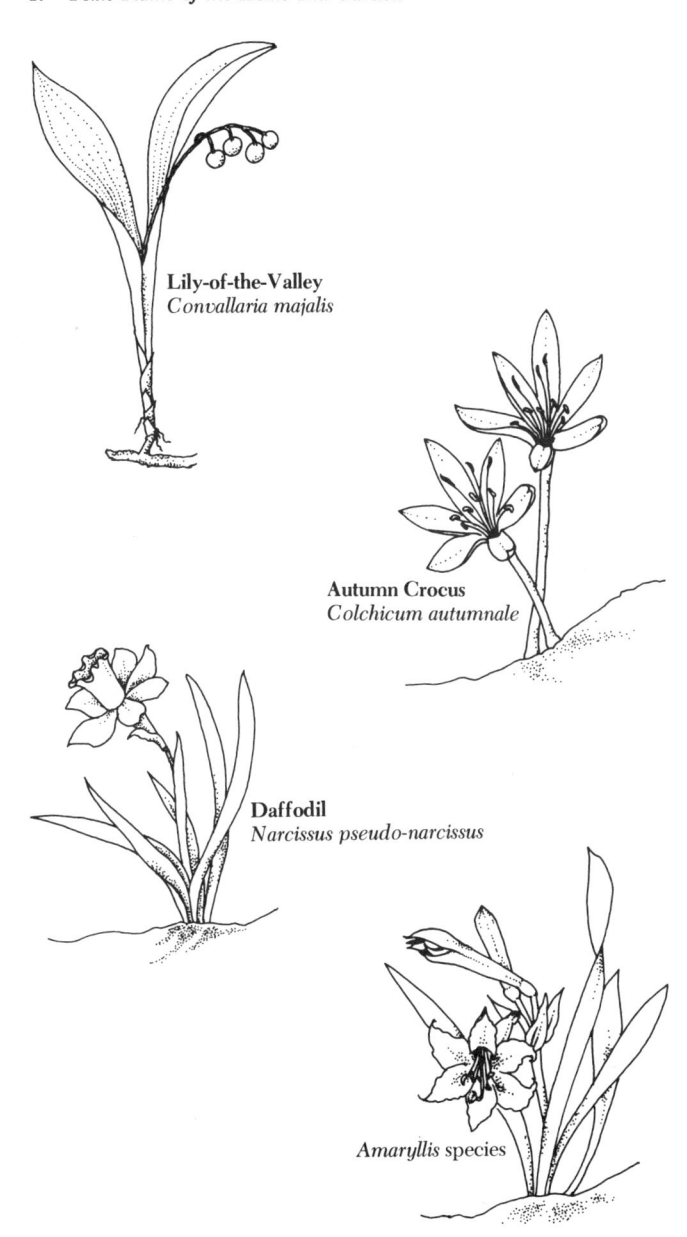

Lily-of-the-Valley
Convallaria majalis

Autumn Crocus
Colchicum autumnale

Daffodil
Narcissus pseudo-narcissus

Amaryllis species

plants planted in the garden or growing wild with the edible plants. In particular, be very careful in your identification of a bulb-forming wild onion, because you may have picked one of its potentially deadly relatives.

LILY-of-the-VALLEY (*Convallaria majalis*)
(all parts, particularly the berries) rarely ☠

This small herb is one of the delights of spring, with its lovely stalk of bell-shaped drooping flowers and delightful fresh fragrance. The flower stalk is produced between two or three oval leaves, each about six inches long, which may be curled in at the margins. The herb lives for many years, deriving its longevity from its deep, creeping underground rootstalk. In the summer, small translucent red berries with several seeds are produced, which may persist long after the leaves have wilted. (**Plate 6.**)

Distribution and Habitat: The lily-of-the-valley is a native of Eurasia but is commonly planted throughout North America in shady gardens, where it may persist long after the garden has been abandoned. For this reason, the plant is often found near abandoned homesites in woods. The native species, *C. montana*, is found on rocky slopes and in open woods in the southern Appalachians from Virginia to northern Georgia.

Toxins: All plant parts contain cardiac glycosides with digitalis-like activity as well as saponins. The poison is not only found in the plant, but may also enter the water in which a bouquet of flowers is being kept. The most common source of poisoning seems to be from children eating the red berries. Considerable research has focused on the cardiac glycosides for use in treating heart disease. However, while herbalists may be tempted to experiment with extracts of this plant, the severity of the symptoms indicates that this is a potent plant best left to medical researchers for the present time.

Symptoms: The saponin may act as a local irritant, but the real danger in this plant comes from the cardiac glycosides which act on the heart. After eating a plant part, the victim may suffer from nausea and vomiting, with some sharp abdominal pain. The pulse may slow down and become erratic. In severe cases, blood pressure may rise. Eventually, the heart ceases to function and the victim dies.

Treatment: See p. 6 for treatment of *digitalis*-like alkaloids.

49 Toxic Plants of the Home and Garden

DAFFODIL, *(Narcissus)* (bulbs) rarely ☠

The Greeks considered the *Narcissus* flower so beautiful that it was incorporated in their mythology: A beautiful boy was so in love with himself that he gazed endlessly at his own reflection in a pond until the gods turned him into a *Narcissus* flower, hence the word "narcissisum." The daffodil *(Narcissus pseudo-narcissus)* and related species such as jonquils are widely planted in gardens and used as potted plants. The most common varieties have flowers that are bright, sunny yellow and trumpet-shaped, though varieties can be found that are orange or white, such as the paperwhite *Narcissus*. The fruit is a green capsule. The long, flattened leaves and perennial bulb look similar to onions, and cases of poisoning occur through mistaken identification. (**Plate 7.**)

Toxins: Daffodils and related bulb-forming genera of the *Amaryllis* family such as *Amaryllis, Hippeastrum,* the *Crinum* lily, the blood lily *Haemanthus,* the *Nerine* lily, the snowdrop *(Galanthus nivalis),* and the Atamasco lily *(Zephyranthes atamasco),* contain toxic alkaloids, such as narcissin, narcipoetin, and lycorin and certain bitter substances. It would make good sense to have your bulb-forming flowers planted well away from your onion patch. You had better be sure that the wild onions in your gazpacho are really onions and not a wild and deadly bulb of something else.

Symptoms: After eating the plant, in particular the bulb, the victim may experience difficulty in swallowing, as well as nausea, vomiting, diarrhea, sweating, trembling and even convulsions. Severe cases may be fatal.

First Aid and Medical Treatment: Induce vomiting and administer activated charcoal and cathartics. Treatment is mainly supportive.

CLIMBING LILY (*Gloriosa* sp.) (tubers) rarely ☠

This graceful climbing plant, a native of tropical Asia and Africa, is now planted outdoors throughout the tropical and subtropical Americas and is also grown as a potted plant. The plant is readily recognized by the long, tendril-like tips of the leaves, which twine around other plants and latticework. The oval to lanceolate leaves themselves are often produced in pairs along the climbing stem. The large flowers are on long stalks arising from the upper leaves. The flowers look like tropical butterflies, with their six yellow to red undulating petals pointed upward. The six stamens are directed downward almost like little legs. The fruit itself is a small capsule containing red seeds, and the plant has a tuberous rootstock.

Toxins: All plant parts contain alkaloids related to colchicine, with the tubers containing the greatest concentrations.

Symptoms: After eating the plant, victims may experience either a burning sensation or a numbness in the mouth and throat, associated with vomiting, stomach cramps and diarrhea. The victims may experience respiratory difficulties and excitation of the nervous system, including convulsions, leading to death within four hours in severe cases.

First Aid and Medical Treatment: Empty gastrointestinal contents and administer activated charcoal and cathartics (see p. 6). Treat other symptoms as they appear.

AUTUMN CROCUS, MEADOW SAFFRON
(*Colchicum autumnale*) (bulb) rarely ☠

Colchicum was known in ancient Greek and Roman times for its medical and toxic properties. The original plants were obtained from the Colchis region around the Black Sea, from which the plant derives its name. *Colchicum* is now widely cultivated across North America for its bright pink or white crocus-like flowers that are produced from the ground in the fall, after the one-foot-long tufted grass-like leaves have already withered away. The fruit matures the following spring into a capsule with many seeds. The plant itself is a perennial, having an onion-like bulb about three inches wide. (**Plate 8.**)

Toxins: The toxic principle is the alkaloid colchicine and related compounds. Colchicine is found in all plant parts, and may be lethal in doses as low as 20 mg, the amount contained in five seeds. Colchicine acts to inhibit spindle formation during normal cell division. The former medical use for treatment of rheumatism and gout and the powerful toxic effects of this plant may be related to this cellular disruption. Colchicine is widely used in genetic research to deliberately alter cell division and create new plants and even animals that have extra sets of chromosomes. These new plants with more chromosomes often have bigger fruits and leaves than normal plants, leading to higher yields.

Symptoms: Two to six hours after eating the plant or drinking milk from sheep or goats that ate the plant, the victim experiences a burning sensation in the mouth, vomiting, diarrhea, severe stomach cramps and difficulty in swallowing. Kidney damage or failure may be evident in blood-tinged urine. An overall general weakness may ensue, leading to circulatory and respiratory failure and even death.

First Aid and Medical Treatment: If the patient hasn't vomited, induce vomiting and give activated charcoal. Be alert for kidney damage. Monitor EKG and respiration and treat symptoms as they appear.

ASPARAGUS *(Asparagus officinalis)*

Description: This familiar perennial garden vegetable, which belongs to the lily family, is recognized by its thick stalks, which emerge in the spring from a perennial rootstock. The leaves are triangular, brown and scalelike; the branches are modified into flattened photosynthetic surfaces. The small flowers are yellow-green and dangle down. Asparagus plants are either "male" or "female," with the "female" plants producing red berries.

Distribution and Habitat: Asparagus is a widespread garden fugitive along fences and roads and in fields, particularly in loose sandy and loamy soils. It is found in every part of the United States except where it is very hot. California and New Jersey are the leaders in commercial production.

Symptoms: From one to 14 hours after eating asparagus, a distinctive sulfurous odor may be produced in the urine. Eating uncooked green shoots has caused some people to develop skin irritations and blisters; even handling young asparagus stems can cause dermatitis. Scarlet berries on the mature plants attract children, and if eaten can trigger allergic reactions in sensitive individuals. Balancing these potential problems against the exquisite flavor of asparagus, let us continue to consume this tasty vegetable (cooked!).

MUSTARD FAMILY

Members of the mustard family (the Cruciferae or Brassicaceae) provide the spice and tang of our food. What would a baseball game be like if one didn't have mustard to put on the hot dog? Or an Irish dinner without cabbage? Other edibles in the family include some of our best vegetables, many of which are tolerant of cool weather and even some frost, such as cabbage, turnips, broccoli, kale, mustard greens, radish, watercress, cauliflower,

52 Poisonous Plants and Mushrooms of North America

Brussels sprouts, rutabaga, and the king of the Passover plate—horseradish. Numerous wild species are consumed by foragers. The spicy, burning taste which makes mustards so tasty is a warning for insects to back off, that the plant contains some toxic substance. In small quantities, our big human bodies can handle the poison, so we continue to eat foods flavored with them. The garden-variety vegetables need no description, but we can briefly characterize most wild members of the family as being herbs with alternate, deeply lobed, or divided leaves. The flowers are often yellow or white, with four petals and six stamens, an odd and yet distinctive combination. The fruit is usually a dry capsule.

Toxins: Mustards produce a mixture of glucosides which interfere with iodide metabolism. In large and prolonged doses, these toxins can lead to an inhibition of thyroid gland activity. In severe cases, an enlargement of the thyroid, known as goiter, may result, giving the victim a swollen neck. The sharp, spicy taste of mustards is caused by blistering, irritant oils which are glucoside isothiocyanates. Interestingly, these oils which developed in plants to protect them from attack by insects are used by certain specialized insects to locate the plants by smell. And for us humans, we eat mustards just because of the spicy, irritating flavor; perhaps it's just like desiring more ardently the lover who tries to end the relationship.

Symptoms: After eating large amounts of raw mustards, a person may experience vomiting and diarrhea, with blood in severe cases. Prolonged eating of raw mustards, such as cabbage, in large quantities can lead to disruption of iodine metabolism and goiter, enlargement of the thyroid gland in the neck.

First Aid and Medical Treatment: No specific treatment unless vomiting has been severe, in which case provide electrolyte replacement.

CASTOR BEAN PLANT (seeds) deadly ☠

The castor bean (*Ricinus communis*) is commonly planted throughout the southern United States and tropical regions for its bright, very large, attractive, green, purple and even red leaves, some of which sport spots or contrasting colored veins. The alternating leaves are palmately lobed and have a long leaf stalk. The tall, thick stem, usually unbranched, may reach 12 feet or more in height. While in most of the United States this plant is a large annual, in the more tropical areas the castor bean plant develops into a woody shrub. Its flowers are small and relatively

Castor Bean Plant
Ricinus communis

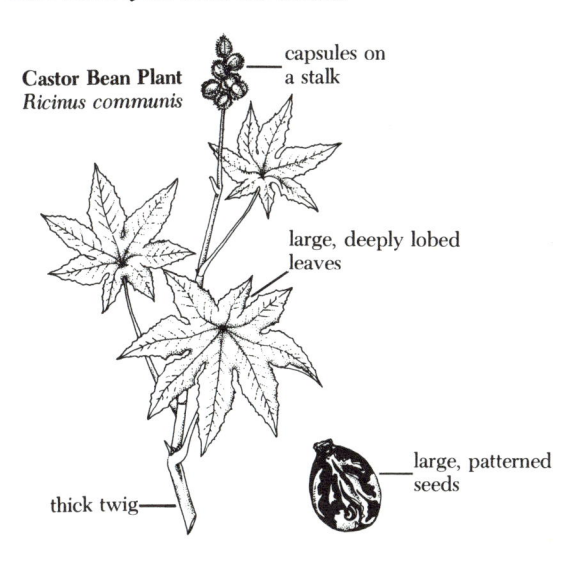

- capsules on a stalk
- large, deeply lobed leaves
- large, patterned seeds
- thick twig

inconspicuous, but the spiny fruits are prominently displayed at the top of the stem. The round, walnut-sized fruits may be brightly colored like the leaves while developing, but are brown when ripe. These fruits split open to reveal three shiny, ellipsoidal seeds which have a mottled coloring of gray, brown, black or white.

This African native is planted throughout North America as an ornamental, but is more common in warmer areas. It has escaped from cultivation and now grows as a weed on disturbed ground. The castor bean has been used since ancient times by the Egyptians as a body oil and a cathartic. Today, castor bean oil is widely used as an industrial lubricant, in paint, leather finishing and fabric treatment. Production in North America is centered in the south central states and California. Castor beans have been used in folk medicine as a cathartic to relieve constipation, sometimes with success and sometimes with fatal results. Tourists shopping in Mexico should beware of necklaces strung with castor beans, which can be dangerous if sucked on or chewed. (**Plate 9.**)

Toxins: All parts of the plant contain the phytotoxin ricin, one of the most deadly substances known. The concentration of ricin is greatest in the seeds. Even eating one seed may be fatal to a child; eating three seeds may kill an adult. The seed coat has to be broken open to release the ricin, so eating a whole seed may not necessarily result in poisoning. Heat apparently destroys ricin, however, other heat-stable poisons in the plant may still induce sickness. Ricin is so lethal that it is estimated that an injection of 0.01 mg is sufficient to kill an adult, which is why this poison was chosen for the recent assassination (by a weapon disguised as an umbrella which injected a pellet of ricin) of a highly vocal BBC broadcaster who criticized the Bulgarian government. The poison is a complex protein, which appears to act by causing a massive allergic reaction against normal proteins, leading to tissue damage in the victim's body. The poison is not soluble in castor bean oil, so that products made with this oil are not necessarily dangerous.

Symptoms: Many hours after eating any plant part, in particular the seeds, a burning of the mouth and throat may develop, followed by nausea, vomiting, diarrhea, abdominal pains, and extreme thirst. Symptoms include blurred vision, general weakness and jaundice, reflecting gastrointestinal and liver damage. Kidneys and lungs may become swollen and damaged as the result of cell breakdown. Death may occur within several days, but even non-fatal cases may cause permanent organ damage.

First Aid and Medical Treatment: If patient is alert and has not vomited, induce vomiting, followed by activated charcoal and laxatives. Be alert for evidence of kidney damage and maintain electrolyte balance.

Blood transfusions may be required and the urine should be kept alkaline.

FOXGLOVE (all parts) rarely ☠

"The Foxglove's leaves, with caution given,
Another proof of favouring heav'n
Will happily display;

The rapid pulse it can abate,
The hectic flush can moderate
And, blest by Him whose will is fate,
May give a lengthened day."

W. Withering, 1785

Two hundred years ago, the foxglove was identified by the English doctor Withering as an important herb in the treatment of heart disease. The drug digitalis, one of the most useful drugs in cardiac treatment, is obtained from the foxglove. The life-saving drug can become a deadly poison when parts of the foxglove are eaten or too much digitalis is prescribed. Poisoning by foxglove can occur because this European native is widely planted in gar-

Foxglove
Digitalis purpurea

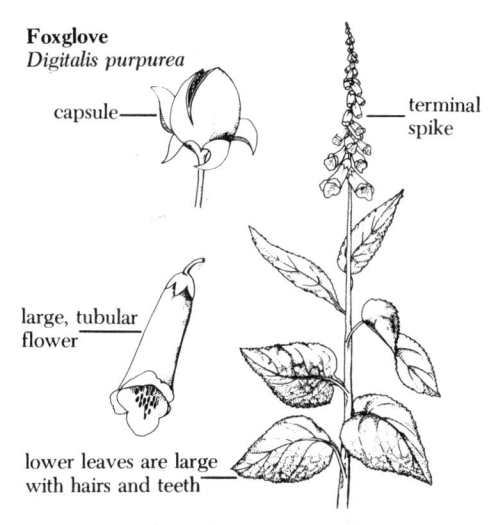

- capsule
- terminal spike
- large, tubular flower
- lower leaves are large with hairs and teeth

dens for its exquisite three-foot-tall columns of large, hanging, bell-shaped flowers. The three-inch-long flowers may be white, pink or purple, often with dark spots inside. The plant grows as a leafy biennial, only flowering in the second year, after which the plant dies. The bottom leaves are large and hairy, with marginal teeth. Foxglove has escaped from cultivation throughout North America and is found commonly in disturbed and logged land in the Pacific Northwest. (see cover, upper left)

Symptoms: While there are very few instances of recorded foxglove ingestion, there have been cases of people poisoned when they mistook the large, soft leaves for salad greens or by sucking the juice from the flowers. The victims experience inflammation of the mouth, followed by nausea and vomiting, abdominal cramps and diarrhea. Severe poisoning, though unlikely, may produce dulling of vision, slowing of the heartbeat, tremors, convulsions and, rarely, death.

Toxins: All parts of the plant contain an array of cardiac and steroid glycosides, poisons which are attached to sugar molecules and activated after they are eaten. The most active component appears to be digitoxis, used in the drug digitalis. In very low concentrations, digitoxin has important medicinal value in cases of a weak heart and irregular, rapid heartbeat. By increasing calcium movement in heart muscle cells and improving muscle tone, digitoxin causes a stronger, more regular heartbeat. However, at higher concentrations, digitoxin apparently reduces the ability of the cell membranes to pump calcium, potassium and sodium in and out of the cell interior, leading to a gradual reduction in function in body cells generally and heart muscle cells in particular.

First Aid and Medical Treatment: Treat for *Digitalis* poisoning, see p. 6.

BEAN FAMILY

One tends to think of the bean family as a real friend to people because of its many useful species like peanuts, clover, alfalfa, and beans. However, the bean family, called both the Leguminosae and the Fabaceae, has many species that can sometimes be dangerous, including such important food crops as bush beans and pole beans *(Phaseolus vulgaris)* and fava beans *(Vicia faba)*, and ornamental plants such as *Wisteria,* rosary pea *(Abrus precatorius),* lupines, hyacinth bean *(Dolichos lablab),* Kentucky coffee tree *(Gymnocladus dioica),* golden chain *(Laburnum anagyroides),* sweet pea *(Lathyrus)* and poinciana *(Poinciana gilliesii).* Poisonous wild members of the bean family include the black locust trees *(Robinia pseudoacacia),* the mescal bean *(Sophora secundiflora)* and herbs known as sesbane, rattlebox and bladder pod *(Crotalaria, Daubentonia, Glottidium,* and *Sesbania).* This abundance of poisonous plants in the Bean Family can be understood in terms of their natural history. In most plants, the element nitrogen, which is an essential component of protein and DNA, is in short supply. Species in the bean family have overcome this problem by making nodules (little round balls) in their roots which provide homes for special bacteria that can absorb nitrogen directly from the air. These bacteria release excess nitrogen into the plant roots. Plants of the bean family then use this abundant nitrogen to produce leaves and seeds with an unusually high nitrogen content. But just like the millionaire who must hire a watchman to guard his property, these plants must protect this treasure of nitrogen by producing a variety of poisons to prevent their being eaten. Their chemical defenses that have been documented include alkaloids, phytotoxins, saponins and cyanide.

Members of the bean family can be readily recognized by a few distinct characteristics. First, the relatively large seeds are contained in a long bean pod; for example, think of pea pods and string beans. The pod is usually green when young, becoming dry and brown at maturity and then splitting to release the seeds. The leaves are produced alternately along the stem, and are usually divided up into three or more leaflets. The five-petaled flowers may be almost any color, but are usually bilaterally symmetrical with one petal larger than the four others.

Most of the poisonous members of the bean family cause gastrointestinal distress, which can be treated by inducing vomiting and washing out the stomach. More dangerous symptoms develop following the eating of certain species, such as the rosary beans. Most cases of poisoning occur through children eating the young bean pods and mature seeds, mistakenly believing that they should be just as edible as the garden varieties of the bean family. Among the most poisonous species of the family are the following:

ROSARY PEA, PRECATORY BEAN, CRABS-EYE, JEQUERITY, PRAYER BEAN, LOVE BEAN (seeds) very dangerous ☠

The rosary pea *(Abrus precatorius)* accounts for the majority of plant fatalities in the subtropical United States. The parts of the plant most commonly encountered are the small pea-sized seeds, which are brilliantly scarlet-colored with a large black spot at one end. The seeds, produced in a hairy pod, are hard and highly shiny, as if painted with enamel. The plant itself is a perennial vine, climbing to 20 feet, with about 12 leaflets on each compound leaf. The flowers are red to purple. (**Plate 12.**)

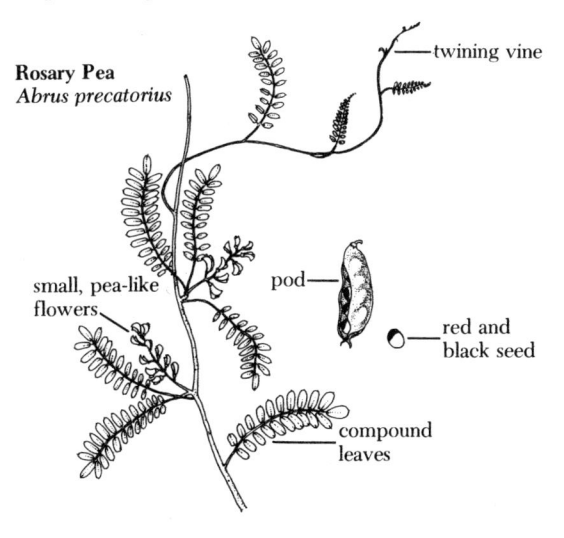

Distribution and Habitat: The rosary pea is a widespread vine of the tropics. The plant has been introduced as an ornamental vine into south Florida, from which it has escaped as a weed. The seeds are so unusually attractive that they are used to make necklaces, rosary beads, toys and other ornaments. Tourists visiting Mexico, Central America, the Caribbean and Hawaii may purchase these lovely items.

Toxins: Even one seed contains enough of the phytotoxin abrin to kill an adult, if the seed is thoroughly chewed up. The lethal dose of abrin is about one part per million in an adult, making it among the most lethal substances known. Abrin consists of two polypetide chains which enter cells in a manner similar to that used by viruses. Inside the cell, the chains inhibit protein synthesis within the ribosomes. This has the effect of causing severe irritation to any exposed mucous membrane, such as the gastrointestinal system. In the past, this powerful effect of abrin on membranes led to limited and often injurious use of this plant to treat eye diseases.

Symptoms: Poisoning may occur through eating the rosary peas from wild plants or sucking on the peas used to make necklaces or other ornaments. Swallowing a single unbroken seed may not be serious, as the hard seed coat may limit leakage of the poison. However, seeds strung on a necklace already have had the seed coat broken; within one hour to several days of ingesting the beans or sucking on them, the victim may experience nausea, vomiting, bloody diarrhea, high fever and stomach pain. Severe symptoms which often do not begin for several days include abnormalities in the heartbeat, seizures, collapse and death. Autopsies may reveal kidney and liver damage.

First Aid and Medical Treatment: If patient has not vomited, induce vomiting. Poisoning should be treated symptomatically, with fluid levels, red blood cell count and electrolyte balance carefully monitored. Even patients showing no symptoms should be very carefully observed, due to the high incidence of fatalities and the often long delay in the onset of either mild or severe symptoms.

LUPINES (seed pods) rarely ☠

Lupines get their name from an old Roman superstition that these plants could attack the soil just like *Lupus*, the wolf, could attack animals. Lupines (*Lupinus* species) include commonly cultivated European garden annuals and perennials that may escape to disturbed habitats. Many American lupines also exist, particularly as shrubs, in western North America. The brilliant yellow, purple,

Lupine
Lupinus perennis

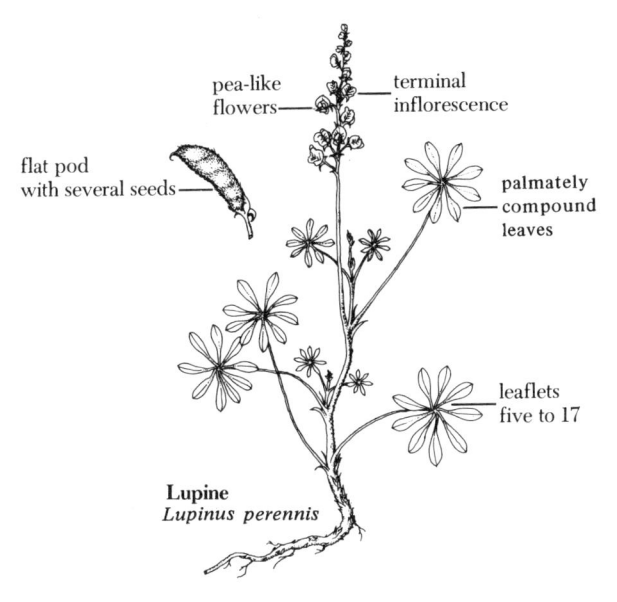

(Labels on figure: pea-like flowers; terminal inflorescence; flat pod with several seeds; palmately compound leaves; leaflets five to 17)

white or blue flowers are produced in cylindrical flower stalks. The pod, containing four to 11 seeds, is about two inches long and is often silky and hairy. The palm-shaped leaves are divided into five to 15 leaflets that are frequently hairy below. (**Plate 13.**)

Toxins: A huge number of alkaloids have been identified from different species. The alkaloids appear to act on the nervous system.

Symptoms: Children may eat the seeds and pods, mistaking them for pea or bean pods. Initial symptoms are vomiting, increased salivation, and general distress, followed by a lowering of the heartbeat and respiration rate, leading to death by respiratory failure in extreme cases of poisoning.

First Aid and Medical Treatment: See Rosary pea, p. 58.

BLACK LOCUST

Black locust trees (*Robinia pseudoacacia*) are recognized by the dark, furrowed bark and the slender, straight trunk. The leaves are divided into seven to 19 small, elliptic leaflets, each about one and a half inches long. A pair of persistent spines is often produced on the twig at the base of the leaf stalk. The fragrant white flowers

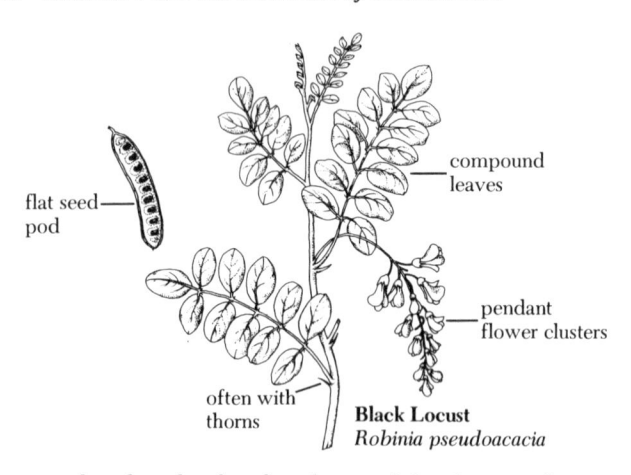

60 Poisonous Plants and Mushrooms of North America

*Labels on illustration: compound leaves; flat seed pod; pendant flower clusters; often with thorns; **Black Locust** Robinia pseudoacacia*

are produced in dangling bunches, and develop into flat, green pods.

Distribution and Habitat: Black locust is native to eastern North America on disturbed ground, along field edges and on sandy soil, but is now widely planted and naturalized throughout North America.

Toxins: The toxic compounds are not fully identified, but apparently include glycosides and phytotoxins.

Symptoms: After eating the seeds, flowers, twigs, or other plant parts, the victim may experience dizziness and depression, associated with diarrhea, vomiting, reduced heartbeat, and poor circulation.

First Aid and Medical Treatment: Induce vomiting and give activated charcoal. Treat symptoms as they develop.

CULTIVATED BEANS

Several commonly cultivated types of beans may cause poisoning. For example, the common kidney bean *(Phaseolus vulgaris)* with its associated varieties, the French bean, the pole bean, Kentucky wonder bean and bush bean, is often eaten as a cooked vegetable. The raw beans and bean pods may contain small amounts of cyanogenic glycosides, which break down to produce cyanide after eating. Cooking makes these foods safe to eat by destroying the glycoside molecule and driving off the cyanide. However, eating large amounts of raw beans can lead to vomiting, diarrhea, and stomach pains. Similar symptoms can arise from eating the raw bean pods of the scarlet runner bean *(P. coccineus)*, which is grown both as an annual ornamental vine and for its edible seeds. While lima beans *(P. lunatus)* grown in North America typically have low cyanide contents (less than one part per 10,000), in certain tropical areas lima beans may have dangerously high cyanide levels (one part per 300). Lima beans with this much cyanide may be poisonous even after cooking. Beans imported into North America are usually inspected to make sure that they have a low cyanide content.

A rare disease of people from Sardinia and the Mediterranean area is now known to be an induced enzyme deficiency caused by eating fava beans, a flat, brown, green, purple, or black bean. This disease of mainly young boys is genetically based and so runs in families. Symptoms such as vomiting, dizziness, headache, stomachache and fever can occur quickly after breathing the pollen or within one day after eating the beans. Symptoms can subside, or in severe cases lead to general anemia, blood in the urine and jaundice lasting from several days to a month.

First Aid and Medical Treatment: See cyanide poisoning, p. 6.

WISTERIA   *(Wisteria sinensis* and *W. floribunda)*

The *Wisteria* is a cultivated vine often trained onto trellises and gazebos, hardy in the north, but most common in the southeastern states, where it has also become naturalized. *Wisteria* has masses of showy, fragrant, sweetpea-like flowers in hanging bunches. The

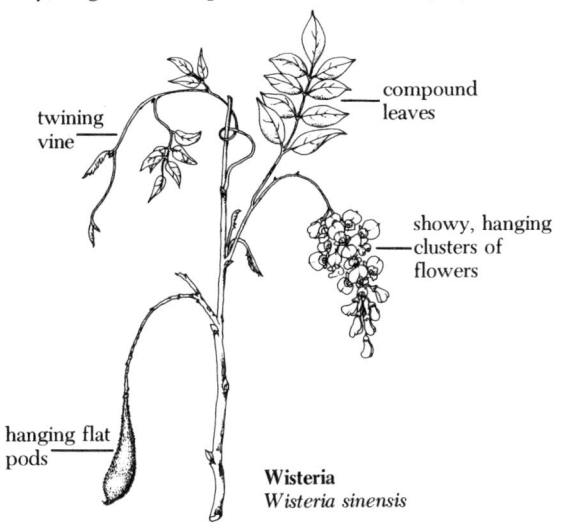

Wisteria
Wisteria sinensis

flowers are generally bluish purple, but there are pink and white varieties. The seeds are contained in thick pods. The alternate leaves are divided into about 11 individual leaflets.

Toxins and Symptoms: All parts of the plant contain an alkaloid of unknown structure. Statements in the older literature that the flowers are nontoxic are in error. Ingestion causes nausea, abdominal pain and repeated vomiting. Diarrhea is slight or absent. Serious intoxications (from chewing the bark) have resulted in sufficient fluid loss from vomiting to result in shock.

First Aid and Medical Treatment: Replacement of fluid losses due to vomiting is an essential part of the management. The administration of anti-emetic drugs may be useful. The victim usually recovers within 24 hours.

THE NIGHTSHADE FAMILY

Eating many species of the nightshade genus *Solanum* can turn your day into a long and permanent night. The scientific name itself comes from the Latin *solanem* or quieting, because of the ancient use of nightshades as sedatives. As Friar Laurence said to the love-stricken Juliet, "And this distilled liquor drink thou of. When presently through all thy veins shall run a cold and drowsy humor; for no pulse shall keep his native progress, but surcease. No warmth, no breath, shall testify thou livest." The curious aspect of the nightshade family is that, beside containing some of our most dangerous wild plants, the family also contains some of our most useful edible plants, such as the tomato, the bell pepper, the chile pepper, the eggplant and the pepino. The potato is a mixed example because the tubers are edible, but all other plant parts are quite poisonous. The nightshades are a large and diverse genus of mainly tropical and temperate plants containing more than 1,000 species, varying in growth form from herbs to shrubs, vines and even trees. Nightshades are often hairy or spiny, but many are also smooth. The leaves are typically alternate along the stem, but can vary greatly in form. The distinctive feature of nightshades is the star- or bell-shaped flower with fused petals, and the stamens joined together to form a distinct yellow beak at the front of the flower surrounding the style. The fruit is usually a pea- to grape-sized fleshy berry with many small seeds. The following are the most important species, though all species of nightshades and other members of the nightshade family, Solanaceae, should be treated as potentially dangerous.

Toxins and Symptoms: All members of this genus contain a glycoalkaloid called solanine. Solanine is most concentrated in green, unripe fruits and in the young leaves and shoots. While adults experience only mild symptoms with solanine ingestion, there have been cases of fatalities in children. Solanine intoxication initially mimics bacterial gastroenteritis and thus is often not diagnosed correctly. Symptoms, which appear a few hours after ingestion, include a scratchy feeling in the throat, nausea, an elevated temperature and diarrhea.

First Aid and Medical Treatment: If patient hasn't vomited, induce vomiting. Treatment is generally symptomatic for gastroenteritis. Compensate for severe fluid losses with fluid electrolyte replacement.

POTATO, IRISH POTATO

Under cultivation since ancient times in the Andes of South America, the potato, *S. tuberosum*, is now cultivated throughout the world and is the single most productive crop in the world on the basis of sheer weight. Curiously, it was the failure of the potato crop in the 1840s that led to the mass immigration of Irish to the Northeast, radically changing American political, social, and economic life. Potatoes may be quite familiar to supermarket shoppers, but the plants themselves are relatively unknown to many urbanites. Potato plants look very similar to tomato plants, with one-to three-foot-long weak or sprawling stems bearing four- to ten-inch-long leaves divided into about nine large leaflets and many smaller leaflets. The flowers are white with yellow stamen, developing into grape-sized, round fruit, yellow to green in color. Poisons are contained in the fruit and any green part of the plant, such as the leaves, shoots, green parts of the potato and sprouts coming out of the potato. Occasionally, in plant supply catalogues, tomato plant tops grafted onto potato plant rootstalks are advertised as a wonder plant capable of producing both tomatoes and potatoes at the same time. These gimmicky plants should be avoided, since any potato leaves developing on the rootstalk will produce poison that will enter the developing tomatoes.

WOODY NIGHTSHADE, CLIMBING NIGHTSHADE

(berries) rarely ☠

The woody nightshade, *S. dulcamara*, is a perennial, climbing or sprawling vine with a slender stem that is native to the Old World but is now widely naturalized, most commonly in eastern North

blue flowers
stamens forming a beak
leaves variable in shape; often with lobes
reddish-orange fruit with many seeds
stem sprawls onto other plants

Climbing Nightshade
Solanum dulcamara

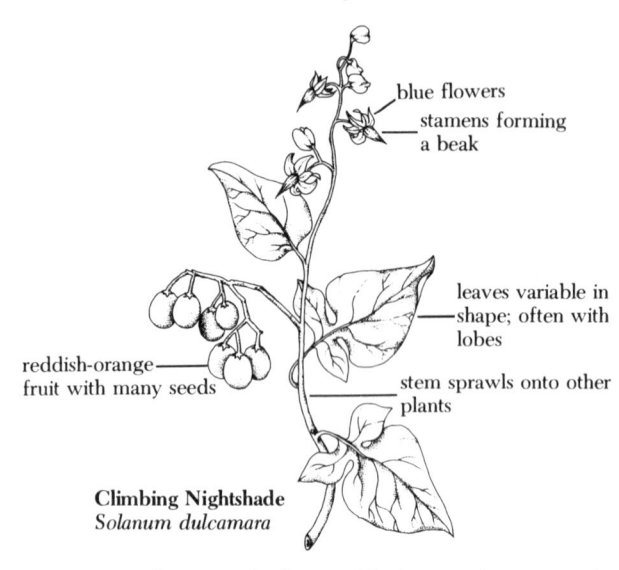

America in weedy areas, thickets and hedges, and on river edges. The leaves are two to four inches long and vary considerably in shape from lobed to oval to heart-shaped. The flowers are white to violet, developing into bright, orange to red, egg-shaped, hanging clusters of berries about the size of grapes. A child could easily imagine these nightshade fruits are like cherry tomatoes, and start eating them. Unfortunately, the berries and probably the rest of the plant are strongly poisonous. (see cover, lower left and **Plate 14**.)

DEADLY NIGHTSHADE, BLACK NIGHTSHADE (berries) rarely

Deadly nightshades, including both a native species (*S. americanum*) and a European species (*S. nigrum*), are annual herbs that occur as weeds in gardens, roadsides, fields and open woodlands throughout North America. The ovate, often slightly toothed leaves are produced on a branched, erect or sprawling stem. The small, white flowers are star-shaped, and develop into hanging, dull to glossy, black fruit. The fruits and other parts are highly poisonous. Certain varieties of deadly nightshade have been developed, with a large production of edible berries and edible leaves. These varieties are usually sold under the name "wonderberry," "sunberry," or "garden huckleberry" and are sometimes considered as a separate species, *S. burbankii, S. intrusum*

65 Toxic Plants of the Home and Garden

Deadly Nightshade
Solanum nigrum

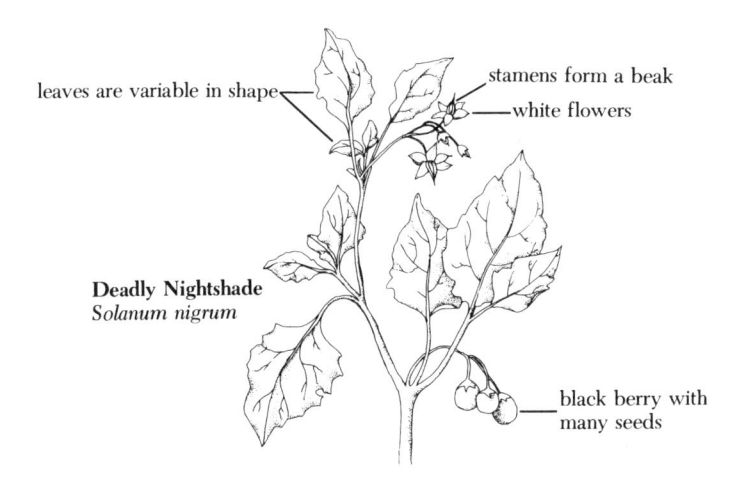

- leaves are variable in shape
- stamens form a beak
- white flowers
- black berry with many seeds

or *S. melanocerasum*. Obviously, one must be careful not to confuse the edible nightshades and the deadly nightshades.

HORSE NETTLE, WILD TOMATO   (berries) rarely ☠

The horse nettle (*S. carolinense*) is a common perennial weed of cultivated fields and disturbed ground throughout North America, but most commonly in the southeastern United States. This herb

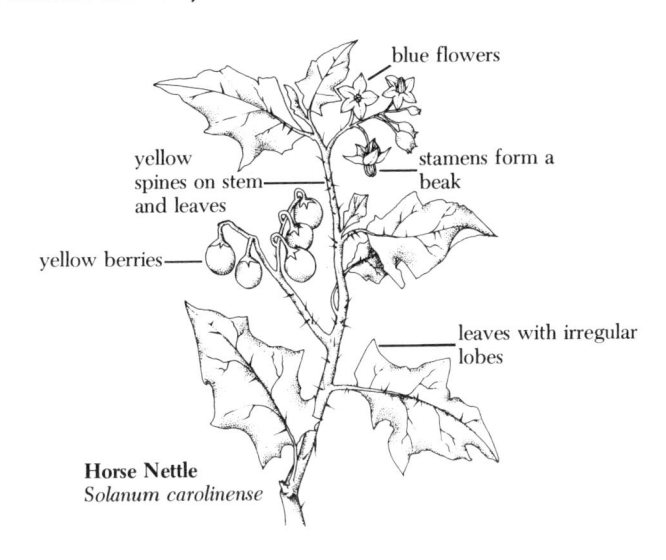

- blue flowers
- yellow spines on stem and leaves
- stamens form a beak
- yellow berries
- leaves with irregular lobes

Horse Nettle
Solanum carolinense

is readily recognized by the yellow spines found on the hairy, erect stem and on the veins of the lobed leaves. Pale blue to white flowers develop into yellow, cherry-sized berries, that often persist in a dried, wrinkled form for many months. These berries have caused the death of at least one child.

JERUSALEM CHERRY (fruits) rarely ☠

The Jerusalem cherry (*S. pseudo-capsicum*) is a delightful potted plant and is also grown as a shrub in Hawaii and other tropical regions. The Jerusalem cherry is grown for the abundant red or yellow cherry-sized fruits that cover the plant and give it a highly decorative appearance. While these fruits look like delicious small tomatoes, they are, in fact, poisonous. The plant has shiny, oblong to lance-shaped leaves and white flowers.

Other potentially poisonous nightshades are the apple of Sodom (*S. sodomeum*) found as a weed in Hawaii and the Devil's apple (*S. aculeatissimum*) which occurs along the southeast coastal plain.

Toxins: Nightshades contain a witch's brew of toxic substances, including the gluco-alkaloid solanine, as well as steroidal alkaloids, saponins, esters, and free parent alkamines. Species vary greatly in the relative amounts and absolute amounts of these poisons, with considerable variation also found over the range of each species. Solanine appears to cause tissue irritation in the mouth,

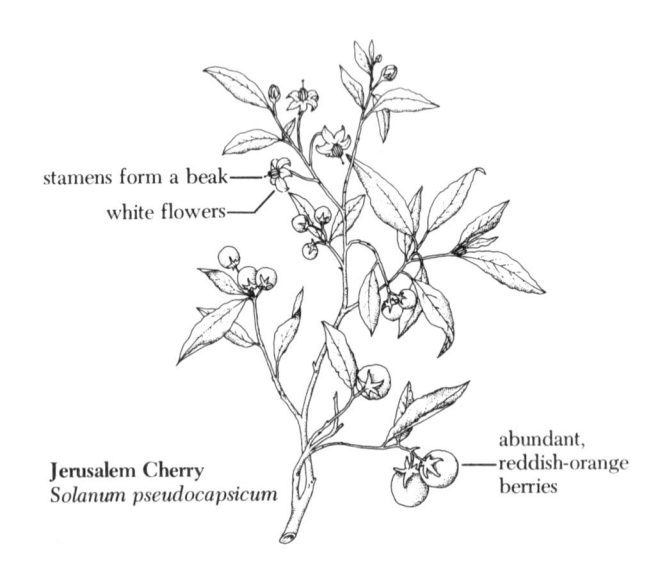

Jerusalem Cherry
Solanum pseudocapsicum

- stamens form a beak
- white flowers
- abundant, reddish-orange berries

67 Toxic Plants of the Home and Garden

gastrointestinal tract and elsewhere, while the steroidal alkaloids are the primary agent acting on the nervous system.

Symptoms: All plant parts, including the leaves, may be poisonous, with the unripe and ripe fruits probably being the most poisonous. The ripe fruits of the woody and deadly nightshades are particularly poisonous, with as few as ten berries potentially being a fatal meal for a child. The first symptoms are an irritation of the mouth and throat, associated with nausea and general and often severe gastrointestinal upset and ulceration. Symptoms associated with effects on the nervous system are increased or depressed pulse rate, dilation of the pupils, loss of sensation, and difficulty in breathing, leading in severe cases to shock, coma, paralysis, and death in rare cases. Generally, the symptoms will disappear within one to two days.

First Aid and Medical Treatment: If patient hasn't vomited, induce vomiting and administer activated charcoal. See Nightshade on p. 63.

OTHER POISONOUS PLANTS IN THE NIGHTSHADE FAMILY

JESSAMINE, *CESTRUM*

Jessamines of the genus *Cestrum* (not to be confused with the poisonous jessamines of the unrelated genus *Gelsemium*) are shrubs with numerous highly fragrant, whitish-green, trumpet-shaped flowers that develop into white or purple berries. Several West Indian jessamines have been introduced as ornamentals into Hawaii and the Gulf Coast from Texas to Florida, from which they have escaped to become naturalized along roadsides. All plant parts are poisonous.

GROUND CHERRY, CHINESE LANTERNS, TOMATILLO, CAPE GOOSEBERRY, STRAWBERRY TOMATO

Ground cherries *(Physalis)* are often grown as ornamental herbs for their orange or purple expanded sepals which give the effect of a hanging lantern. Certain species have edible, yellow or purple, sweet or starchy-tasting berries, within this lantern, which are gathered as food. However, the unripe fruits are toxic.

CHALICE VINE, TRUMPET FLOWER

Chalice vines *(Solandra guttata* and *S. maxima)* are climbing shrubs from Mexico, cultivated in the southern United States for their large, yellowish-orange, lobed, goblet-shaped flowers. Poisoning may result from eating the leaves or flowers.

AROID FAMILY, WITH SPECIAL ATTENTION TO *DIEFFENBACHIA* (all parts)   rarely ☠

Several species of *Dieffenbachia* (dumb cane or dumbcain) and related species in the aroid family (Araceae) probably account for most of the cases of plant poisoning in the home. This is a large plant family, many of which are tropical climbers commonly used as ornamental house plants. These include: the arrowhead-leaved *Anthurium*, the trailing *Philodendron*, the white-flower-stalked *Spathiphyllum* and large-leaved *Monstera* (see montage). Species in the genera *Caladium*, *Xanthosoma*, *Colocasia* and *Alocasia*, commonly known as elephant's ears, are grown as potted plants or outdoors in subtropical regions for their large, arrowhead-shaped or round leaves, often brightly spotted or streaked with white or purple, and for their sometimes edible rootstocks. Also in this family are some common swamp and wet North American woodland herbs such as the odoriferous skunk cabbage *(Symplocarpus foetidus)*, the handsome jack-in-the-pulpit *(Arisaema triphyllum)*, the arrow-arum *(Peltandra virginica)* and the golden club *(Orontium aquaticum)*. The unique calla lilies of South Africa of the genus *Zantedeschia* are aroids. (**Plates 10 and 11.**)

Description: *Dieffenbachias* look somewhat like small banana plants, one to six feet tall. Typically, a plant has a single erect stem. The thick green stem is circled with evenly spaced leaf scars and topped with a cluster of handsome streaked leaves. Leaves are large, oblong, thin, hairless, untoothed green blades, often streaked or spotted with white or yellow. The leaf stalk surrounds the stem. The plant rarely flowers or fruits indoors and is usually propagated through stem cuttings. In general, when aroids do flower, they produce a small spike of flowers surrounded by a large modified leaf that may be green or white. The resulting fruits are often bright red or orange.

Distribution and Habitat: *Dieffenbachias* are commonly grown as ornamentals in houses, lobbies and restaurants and may be grown outdoors as a perennial in south Florida. This species occurs naturally as a perennial understory herb in the rain forests of

69 Toxic Plants of the Home and Garden

Aroids

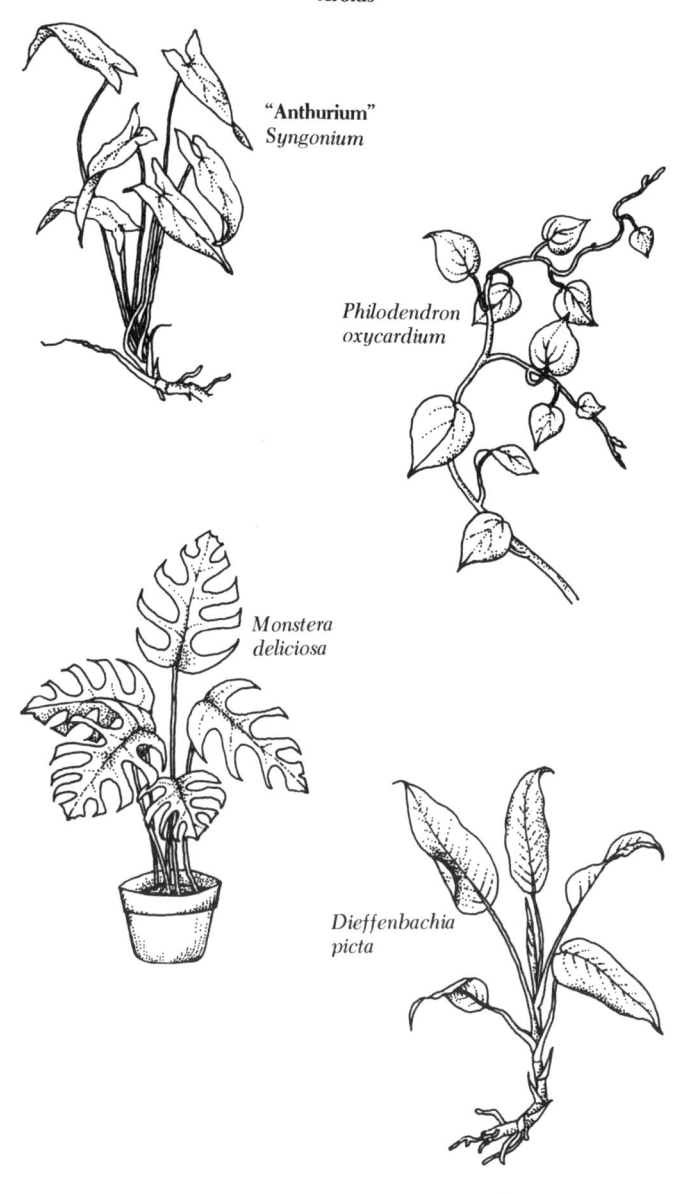

"Anthurium"
Syngonium

Philodendron oxycardium

Monstera deliciosa

Dieffenbachia picta

Skunk Cabbage
Symplocarpus foetidus

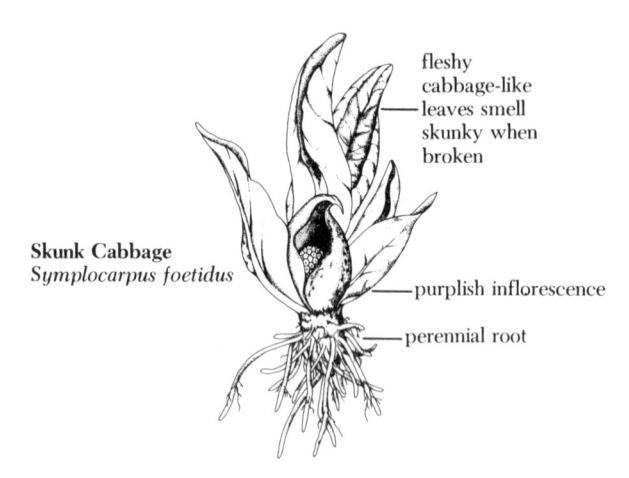

- fleshy cabbage-like leaves smell skunky when broken
- purplish inflorescence
- perennial root

tropical America. The family itself is worldwide, though most abundant in damp and swampy tropical areas.

Toxins: All parts are usually poisonous, although the leaves may sometimes be devoid of poison. The chemical nature of the toxins has been recently attributed to several proteolytic enzymes, which trigger the release of some potent kinins and histamines by the body. These kinins in turn cause several local reactions, which

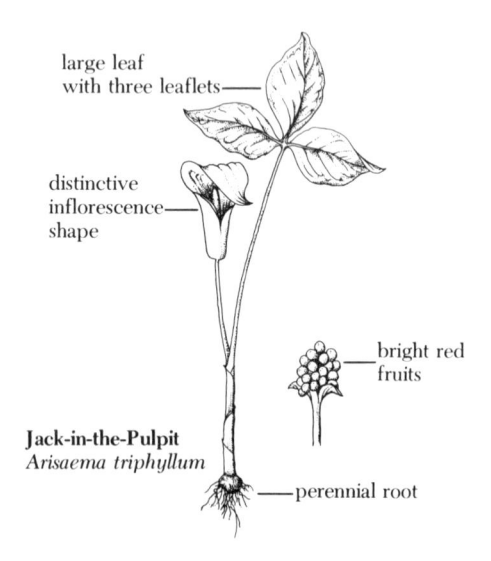

Jack-in-the-Pulpit
Arisaema triphyllum

- large leaf with three leaflets
- distinctive inflorescence shape
- bright red fruits
- perennial root

may be aggravated by sharp microscopic needles (crystals) of calcium oxalate contained in plant tissue. These needles are thought to mechanically damage cells in the mouth during ingestion or may simply allow the poison to enter and create cellular havoc. A recent (1983) study reported that dumb cane has specialized contractile cells that literally act as missile launchers and fire the needle-sharp calcium oxalate spears (rhabdites) into the tissues. Various books on edible plants advise that the leaves and roots of certain aroids, such as skunk cabbage and arrow-arum, can be eaten after being thoroughly boiled or baked. In our experience, some residual poison remains even after cooking, so that edible-plant foragers should be very careful.

Symptoms: Immediately upon chewing plant parts of aroids, an irritating or burning sensation is felt throughout the mouth, tongue, lips, and throat. These symptoms may be followed by copious salivation and swelling of the tongue. Talking, swallowing, and breathing may become difficult. In severe cases, swelling of the tongue can cause choking or death. The difficulty of talking with a swollen tongue accounts for the *Dieffenbachia's* name of dumb cane. Symptoms may subside in several minutes or persist for more than a week. Associated symptoms of nausea, vomiting, and diarrhea may indicate that there are several toxic ingredients in this plant. Sensitive individuals may experience irritated fingers just by picking the flower stalk or handling plant cuttings. In some aroids, severe poisoning may *very rarely* result in irregular heartbeat, dilation of the pupils, fits, and coma, leading to death.

First Aid and Medical Treatment: Normally, symptoms are so immediate that plant tissue is not swallowed. If the plant parts are swallowed and the victim does not vomit, induce vomiting (see p. 5). Treat mouth irritations symptomatically, using soothing mouthwashes of ice-cold water, milk or antacids to dilute or flush out the oxalate crystals. Some investigators report histamine release by dumb cane; hence antihistamines may help relieve the symptoms. Call a physician if the plant parts have caused extensive vomiting. The victim may become dehydrated and will need intravenous fluid replacement. Treat pain with analgesics—meperidine, etc. Aspirin has some effect, inhibiting kinins, but may be hard to swallow with a swollen mouth.

RHUBARB, PIE-PLANT (leaves only are dangerous) rarely ☠

Rhubarb *(Rheum rhaponticum)*, of the buckwheat family, is a favorite easily grown perennial garden vegetable found across the northern United States and Canada and farther south in cooler mountain regions. Its thick, long, often reddish leaf stalks are gathered in the spring and used as a vegetable or as filling for lush, tart rhubarb pies. As early as four thousand years ago, in China, rhubarb leaf stalks were known medicinally for their laxative effect. While the leaf stems are edible, the leaves themselves are toxic and potentially deadly. These heart-shaped leaves are very large, about ten to 17 inches long, with a soft, undulating surface. The leaves look good enough to eat, but are deadly poisonous. The rhubarb plant has a perennial rootstalk which produces the rosette of leaves at ground level in the early spring. If the leaves are not frequently harvested, the plant can form a tall stalk capped with clusters of small, greenish-white flowers.

Toxins: The toxic principle in rhubarb leaves is oxalic acid, a colorless crystal used in industry for bleaching and chemical reduction. It is irritating to any tissue with which it comes in contact. Internally, oxalic acid binds to ions, particularly calcium, to form insoluble crystalline precipitates. These precipitates are dangerous in that the sharp crystals cause cellular and mechanical tissue damage. Oxalates also cause abnormally low levels of blood calcium. The calcium oxalate crystals which accumulate in the kidney can cause serious damage to these critical organs.

Symptoms: Eating the reddish leaf stalks is perfectly safe, whether the rhubarb is cut fresh in the garden or purchased in bundles from the store. However, the inexperienced or extremely frugal gardener may wonder why he shouldn't try eating the lovely succulent rhubarb leaves, particularly as they seem so similar to Swiss chard in appearance. And so our home gardener makes a potentially fatal error. Varieties of Swiss chard with reddish leaf stalks may create additional confusion of identity in the minds of some gardeners. The sap of the broken leaves may cause some degree of irritation in sensitive individuals, but the poison acts primarily when the leaves are eaten. Symptoms usually begin after a delay of about one day and include nausea, stomach pain, vomiting and diarrhea, sometimes bloody, due to internal bleeding. Initially, the victim may experience a burning sensation in the mouth and throat, which may then be followed by abnormal electrolyte imbalance and a drop in blood calcium. The calcium oxalate crystals which form may damage the kidneys, leading to a decrease or even cessation of urine production, and the presence of albumin, blood, oxalic acid or protein in the urine. Decreased calcium levels have an adverse affect on the contractility of all muscle tissues, including the heart, as well as disrupting some functions of the nervous system, leading to such symptoms as muscle cramps, muscle twitches, headache, confusion, convulsions, coma, and in rare cases even death.

First Aid and Medical Treatment: If victim hasn't vomited, induce vomiting and give cathartics to empty the gastrointestinal tract. Monitor blood calcium levels and if below normal, intravenous calcium gluconate may be called for. Treat other symptoms as they occur.

Ivy, English Ivy (berries) rarely ☠

The poison which surrounds our academic institutions is not radical dogma or declining standards, but rather the climbing ivy *(Hedera helix)*, a member of the ginseng family. The adaptable ivy is a jack-of-all-trades, growing up the sides of buildings with anchoring aerial roots, spreading as a ground cover, thriving as a house plant and even standing erect as a shrub. While typical plants have dark green, leathery, three-to-five-lobed maple-like leaves, horticultural forms possessing round leaves and long-lobed leaves exist, often with white streaks or even predominately white leaves with green streaks. While many ivy plants seem never to flower, they actually produce inconspicuous whitish to greenish flowers in small round clusters late in the growing season. The persistent fruit is a pea-sized, black, bitter-tasting berry with three to five seeds.

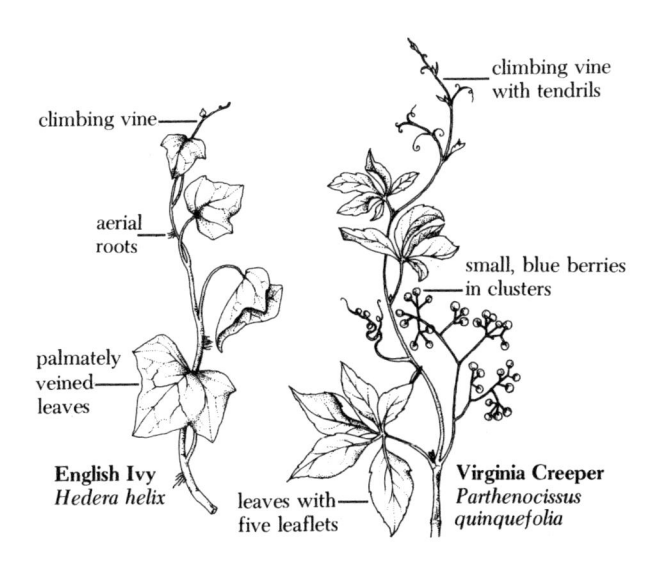

English Ivy
Hedera helix
- climbing vine
- aerial roots
- palmately veined leaves

Virginia Creeper
Parthenocissus quinquefolia
- climbing vine with tendrils
- small, blue berries in clusters
- leaves with five leaflets

Distribution and Habitat: This European native is widely planted as a ground and wall cover, from which it spreads into surrounding habitats. Potted varieties are common indoors.

Toxins: All plant parts contain hederagenin, a triterpenoid saponin compound. The compound appears to react with cholesterol in the cell membranes, resulting in cells bursting.

Symptoms: Poisoning occurs from eating the berries or other plant parts. Children may mistake the berries or grapes for juniper berries. Teas brewed from ivy leaves are also poisonous. Early symptoms are diarrhea and vomiting, followed by excitement, difficulty in breathing, fever, and even convulsions, coma, and rarely death.

First Aid and Medical Treatment: Induce vomiting and administer activated charcoal; otherwise treat symptoms as they occur.

Related species: Other members of the ginseng family, Araliaceae, should be regarded cautiously. The Devil's walking stick or Hercules' club *(Aralia spinosa)*, an uncommon spiny shrub of the eastern United States and related *Aralia* species known as sarsaparilla and spikenard are regarded as having poisonous ripe or unripe berries. American ginseng *(Panax quinquefolius)* is a minor agricultural and wild crop, being exported to East Asia, where it is prized as a sexual stimulant and a general health tonic for aging men. Another widespread climbing vine with poisonous small blue fruits is the Virginia creeper *(Parthenocissus quinquefolia)*, which is in the grape family.

YEW, GROUND HEMLOCK (seeds in red berries) dangerous

Yews are handsome evergreen shrubs or trees up to 75 feet tall, often living to a great age. The narrow, stiff, dark green leaves are about half- to an inch-and-a-half long and are often arranged in two ranks on the twigs. The underside of each leaf is light in color and has a prominent midrib. There are separate male and female plants, each of which produces inconspicuous tiny flowers. The female plants produce fleshy, bright red or pink, round "fruits," which are actually cup-like structures that nearly surround the single, large (¼-inch diameter), greenish seed. The sticky, sweet flesh of the cup is often eaten by birds without ill effect. **(Plate 15.)**

Distribution and Habitat: Cultivated introduced yews are widely planted throughout North America as hedges, shrubs and small trees near homes, building and parks. There are also two wild species native to North America. *Taxus brevifolia* occurs in mountain streams and gorges below 7,000 ft. elevation from Alaska through British Columbia, Montana, and California. *Taxus canadensis* occurs in eastern North America. *Taxus floridana* is found in the woods of northwestern Florida.

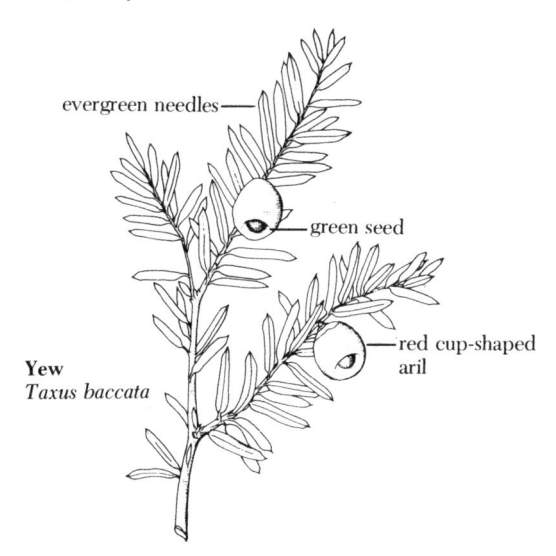

Toxins: A poisonous alkaloid, *taxine*, is found in all plant parts, (leaves, seeds and bark), but the pink flesh of the "fruit" may be free of poison. If an intact seed is swallowed without chewing, poisoning may not occur, but if the seed coat is cracked, even one or two seeds could be fatal for a small child.

Symptoms: Ingestion of taxine, a central nervous system stimulant, after a short delay, causes dizziness and a dry mouth, followed by salivation, diarrhea, vomiting, trembling, pupil dilation, decreased cardiac output, slow pulse, difficulty in breathing, abdominal cramps, and muscular weakness. A rash may appear, the cheeks become pale, and the lips turn blue. In severe cases, collapse, coma, convulsions, and death, which have been attributed to cardiac or respiratory failure, may take place.

First Aid and Medical Treatment: Induce vomiting if victim is alert. Give activated charcoal; give medication for pain if severe and monitor the electrocardiogram. Treatment of symptoms is as indicated.

Prevention: Parents should remove the decorative red berries from all yews to prevent children from eating them.

HORSE CHESTNUT, BUCKEYE (all parts)
(*Aesculus* sp.) rarely ☠

Horse chestnut trees and shrubs are known to all children for their wondrous brown, walnut-sized seed, which looks perfectly stained, waxed and polished. Eager children often pry the large seeds from the spiny green capsules rather than wait for the seeds to fall by themselves. The seeds may be kept as private treasures or used as perfect throwing stones. Horse chestnuts are readily recognized by their large opposite leaves, each leaf divided into five to seven leaflets like the fingers of a hand. The individual leaflets have a toothed margin and have a somewhat crinkled surface. The flowers appear in large, erect, candle-like bouquets, in colors ranging from yellow to white, pink, and red. (**Plate 16.**)

Distribution and Habitat: The European horse chestnut, *Aesculus hippocastanum*, is widely planted for its lovely foliage, growth form and flowering display. Buckeyes are widely distributed native trees and shrubs of rich woodlands and streams in eastern and southern North America and dry hillsides and valleys of the mountains of the Pacific Coast Range and the Sierra Nevada, down into Baja California.

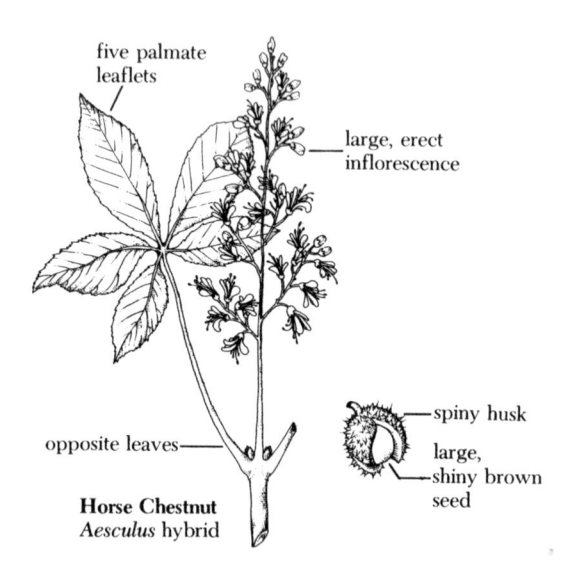

Toxins: All plant parts contain the toxic glycoside aesculin. The poison acts by breaking down blood proteins. In earlier times, people fished by dumping crushed leaves and seeds into ponds to stun fish and cause them to float to the surface, still alive and edible.

Symptoms: Adults or children may ingest the poison by drinking tea made from leaves or eating the elegant seeds. Honey made by bees gathering nectar from the blossoms may also contain the poison. The victim experiences vomiting, diarrhea, and nausea, often associated with headache, stomachache, dilation of the pupils and other visual disturbances, thirst, and mental confusion. In rare cases, poisoning in children can lead to death as a result of respiratory failure.

First Aid and Medical Treatment: If victim has not thrown up and is alert, induce vomiting and administer activated charcoal. Maintain fluid electrolyte balance. Monitor respiration and be prepared to treat respiratory failure. Monitor EKG and be alert for arrhythmias.

CYCADS (all parts) rarely ☠

Fern palm, coontie, Florida arrowroot (species of *Zamia*, *Macrozamia*, *Cycas*, *Dioon*, and other related genera) are striking ornamental plants cultivated throughout the warm areas of the United States and subtropical to tropical regions. Wild species of the Florida arrowroot, *Zamia*, occur in sandy pine savannas of Florida and the Caribbean. Cycads look like lovely small palm trees, with their thick pillar-like unbranched stems and compound, leathery, fern-like leaves. The young leaves unroll just like fern fiddleheads. In fact, cycads are the last living members of an ancient and important gymnosperm family that was outcompeted during the rise of the flowering plants hundreds of millions of years ago. The stem is often covered by the persistent stalks of the old leaves. Cycads lack true flowers and fruits, but bear their seeds in clusters on cone-like structures. The seed is about the size of a walnut, with a thin fleshy layer that becomes yellow, orange or red at maturity.

Symptoms: All parts of the plant may contain poisonous alkaloids. Poisoning may occur through eating the unwashed roots or flesh surrounding the seeds. The roots contain some stored carbohydrates and may be eaten after thorough washing. Instances of poisoning appear to be rare, but reported symptoms include paralysis and even death.

First Aid and Medical Treatment: Induce vomiting and administer activated charcoal. Respond to symptoms as they develop.

78 Poisonous Plants and Mushrooms of North America

CHINABERRY TREE, CHINA TREE, TEXAS UMBRELLA TREE

(leaves, fruit) rarely ☠

The chinaberry tree *(Melia azedarach)* is one of the few lonely members of the large tropical mahogany family that survive in the Temperate Zone. This small- to medium-sized tree may be recognized by the large leaves, one to three feet long, which are divided into small, one-inch-long, toothed leaflets. The overall shape of the leaf is somewhat long and triangular. The pale purple flowers are produced in long-stemmed, open clusters. The yellowish, cherry-sized fruits contain a single seed.

Distribution and Habitat: This native of southwest Asia has been frequently planted as an ornamental tree in the southern United States and Hawaii, but in many cases has escaped to woodland edges and other scrubby areas.

Toxins: A toxic resin, as yet unidentified exactly, appears to be present in the leaves and the fruit. The diversity of symptoms indicates that more than one poison may be present.

Symptoms: Following ingestion of the plant, the victim experiences severe gastrointestinal irritation, including vomiting, diarrhea, and nausea. Symptoms associated with effects on the nervous system are excitement, weak heartbeat, and difficulty in

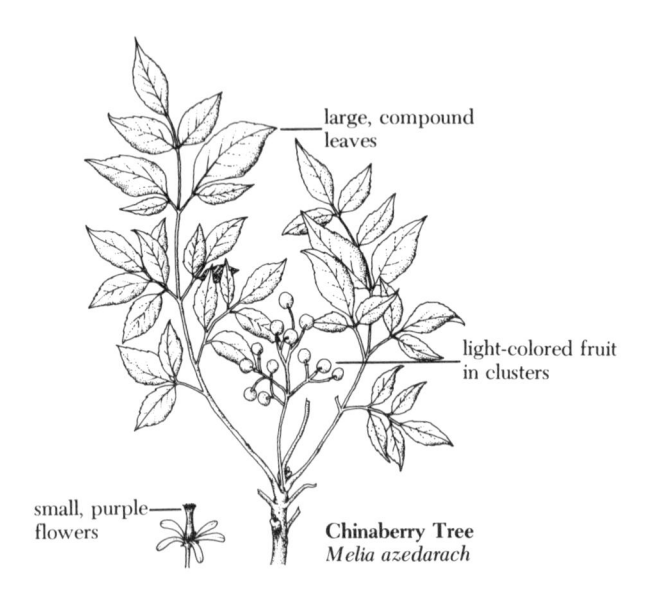

Chinaberry Tree
Melia azedarach

- large, compound leaves
- light-colored fruit in clusters
- small, purple flowers

breathing. Death may occur within one day, with eight fruits being a possibly lethal dose. Liver and kidney damage may occur during the period of poisoning.

First Aid and Medical Treatment: Induce vomiting and administer activated charcoal. Monitor EKG, respiration and kidney function.

SANDBOX TREE

The sandbox tree *(Hura crepitans)* is native to the American tropics, where the bark is sometimes thrown into water to stun fish. This should give one the idea that the tree is to be treated with respect. The sandbox tree is occasionally planted in the south and Hawaii as a curiosity because of the large conical spines covering the trunk and branches, like the doorway into a medieval fort, and the explosive capsules which open with the sound of a rifle shot, scattering the large seeds. The plant apparently contains a mixture of toxins, possibly including a phytotoxin. Contact with the sap or jewelry made from the capsules may cause skin irritation. Eating any plant part may cause vomiting and diarrhea.

JATROPHA, PHYSIC NUT (all parts) rarely ☠

Four tropical American *Jatropha* species are widely planted for their medicinal value and their beauty as ornamentals, often as hedges. Jatrophas have been introduced throughout south Florida, the Gulf Coast and Hawaii. The four most common species are (1) the physic nut or coral plant *(J. multifida)*, (2) the Barbados nut, purge nut, physic nut or curcas bean *(J. curcas)*, (3) the bellyache bush *(J. gossypifolia)* and (4) the peregrina *(J. hastata* or *J. intergerrima)*. Jatropha species are a group of thick-stemmed small trees and shrubs. The plants may be recognized by the alternating relatively large leaves (five to 12 inches wide) with palmate veins, often with three to 13 shallow to deep lobes, and relatively long leaf stalks. The plants give off a sticky yellow sap when cut. The flowers are small, and are either red or yellow. The yellow to brown fruits are round and approximately walnut-sized, with three to six lobes. At maturity, the fruit either splits into three sections, each containing one large, pleasant-tasting brown to black seed, or falls as a unit.

Toxins: All plant parts contain the toxalbumin jatrophin and a purgative oil similar to castor bean oil; moreover, the leaves appear to contain saponin. The toxalbumin severely damages all cells it comes in contact with, often damaging the kidneys once the

80 Poisonous Plants and Mushrooms of North America

Physic Nut
Jatropha curcas

large, palmately veined leaves

large seed

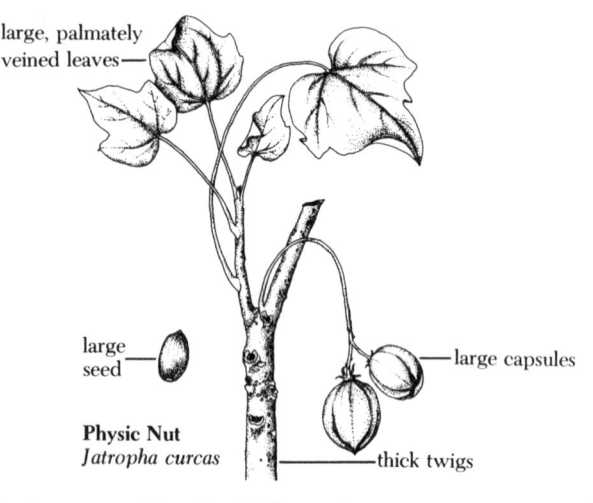

large capsules

thick twigs

toxin has entered the blood. These dynamic properties give the plants such diverse applications as a fish poison and as a folk cure for throat cancer in Aruba.

Symptoms: The most common causes of poisoning occur when people use these plants for medicinal purposes or try out the seeds as food. While the roasting of physic nuts is popularly thought to render them edible or useful as medicine, the seeds may still be toxic. The raw seeds themselves are used as rat poison in Africa, which should be a good tip-off to avoid this plant. Within one hour of eating the seeds or other plant parts, the victim may begin to experience severe gastrointestinal irritation, including a burning feeling in the mouth and throat, diarrhea, vomiting, nausea and stomach pains, followed in severe cases by collapse, convulsions, kidney malfunction leading to blood poisoning, coma and rarely death. The sap of these species is used in the Caribbean to stop bleeding from cuts and wounds, though the sap is irritating to the eye and may cause skin rashes in sensitive individuals.

First Aid and Medical Treatment: Induce vomiting and administer activated charcoal. Monitor for renal damage.

HONEYSUCKLES, ELDERBERRIES, CORAL BERRIES, SNOWBERRIES, VIBURNUMS (all parts) rarely

Members of the honeysuckle family (Caprifoliaceae) are grown as ornamentals and are common small trees and shrubs of our forests. Cases of poisoning by these plants are infrequent, but the common occurrence of these plants and the attractiveness of the fruit make a brief mention appropriate. Potentially poisonous members can be readily identified as always having leaves produced in pairs, a four- to five-lobed, usually white flower tube apparently attached above the ovary, and a fleshy, several seeded fruit, often brightly colored. The twigs are often hollow or filled with a soft pitch, making the twigs attractive to a child for use as a blow-pipe. The elderberries *(Sambucus)* usually have the flowers and edible berries in flat-topped clusters above the leaves and have the leaves divided into three to seven leaflets. Honeysuckles *(Lonicera)* have extremely fragrant flowers, simple leaves, and typically red fruit among the leaves. Snowberries *(Symphoricarpos)* are similar to honeysuckles in appearance, but have white fruit. The genus *Viburnum* is a large and diverse North American group, which includes trees and shrubs known as arrow-wood, withe rod, black haw, nanny bush, cranberry bush, wild raisin, hobblebush, and nannyberry. Some of these *Viburnum* species may have edible berries, but other species have potentially poisonous berries.

Distribution and Habitat: The European fly honeysuckle *(Lonicera xylosteum)*, the climbing woodbine *(L. periclymenum)*, the wayfaring-tree *(V. lantana)* and the guelder rose *(V. opulus)* are European ornamentals with poisonous fruits, and may frequently persist in abandoned gardens or escape into open woodlands. Members of the honeysuckle family are widely distributed across North America in wooded habitats, with the poisonous snowberry *(S. albus)* being a commonly cultivated native species.

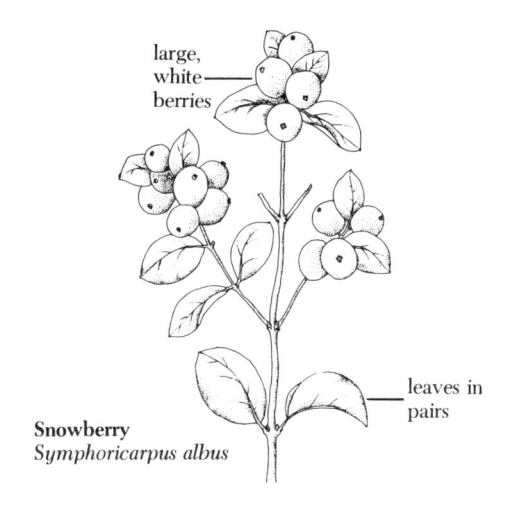

Snowberry
Symphoricarpus albus

Toxins: A variety of toxins including saponins, glycosides, and bitter tannins have been identified as being found in species of this family.

Symptoms: The sap of snowberry fruits has been reported to be irritating to the skin. Most cases of poisoning probably result from eating the brightly colored fruit of poisonous species or using their hollow twigs as pea-shooters. The victim may experience diarrhea, vomiting, stomach cramps and nausea. The symptoms are usually not serious unless the victim has eaten a lot of the berries, particularly of the fly honeysuckle, in which case the victim may experience visual difficulties such as pupil dilation, irregular pulse, sweating and in rare cases even coma and death.

First Aid and Medical Treatment: If patient has not vomited, induce vomiting and administer activated charcoal. Treat symptoms as they appear.

TUNG-OIL TREE, TUNG NUT (seeds) rarely ☠

A major constituent of oil-based paints and varnishes is the oil derived from the Chinese tung-oil tree, *Aleurites fordii*. This small tree has dark green, alternate, ovate to heart-shaped, relatively large (to ten inches long) leaves with palmate veins and a long leaf stalk. Large clusters of white to pink flowers are produced in the spring, followed by a walnut-sized brown fruit containing three to five large, round, rough-surfaced seeds. The fruits

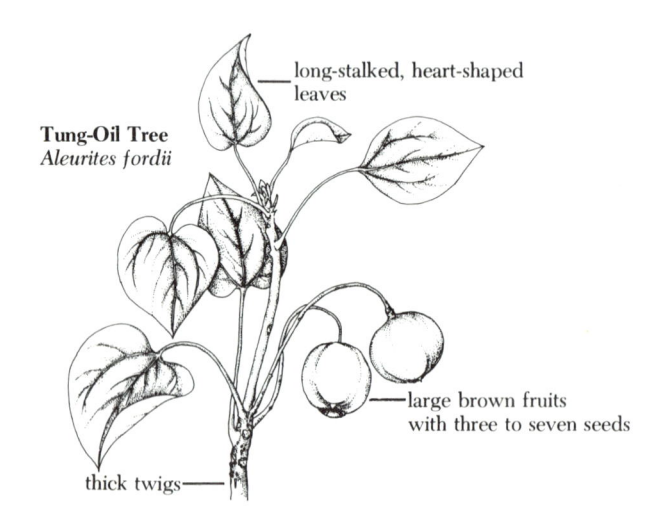

Tung-Oil Tree
Aleurites fordii

— long-stalked, heart-shaped leaves

— large brown fruits with three to seven seeds

thick twigs —

Plate 1. *Stinging Nettles,* p. 25

Plate 2. *Marijuana,* p. 32

Plate 3. *Morning Glory, p. 35*

Plate 4. *Thorn Apple, Jimsonweed, p. 37*

Plate 5. *Opium Poppy, p. 41*

Plate 6. *Lily-of-the-Valley, p. 48*

Plate 7. *Daffodil, p. 49*

Plate 8. *Autumn Crocus, p. 50*

Plate 9. *Castor Bean Plant, p. 52*

Plate 10. *Jack-in-the-Pulpit, p. 68*

Plate 11. *Elephant's Ears*, p. 68

Plate 12. *Rosary Pea*, p. 57

Plate 13. *Lupines, p. 58*

Plate 14. *Climbing Nightshade, p. 63*

Plate 15. *Yew*, p. 74

Plate 16. *Horse Chestnut*, p. 76

Plate 17. *Oleander, p. 84*

Plate 18. *Mountain Laurel, p. 87*

Plate 19. Rhododendron *species*, p. 86

Plate 20. *Mezereon*, p. 88

Plate 21. *Crown-of-Thorns, p. 94*

Plate 22. *Poison Hemlock, p. 101*

Plate 23. *Pokeweed, p. 104*

Plate 24. *Milkweed, p. 107*

Plate 25. *Bittersweet, p. 114*

Plate 26. *May Apple, p. 128*

Plate 27. *Destroying Angel*, p. 136

Plate 28. *Destroying Angel*, p. 136

Plate 29. *Eastern Fly Agaric, p. 144*

Plate 30. *False Morel, p. 141*

Plate 31. Galerina autumnalis, *p. 139*

Plate 32. *False Chanterelle, p. 158*

are held on drooping stalks and generally fall to the ground at maturity. In commercial production, the seeds are gathered, hulled, heated, crushed and pressed to extract the tung oil.

Distribution and Habitat: Large plantations of Chinese tung-oil trees were formerly common along the Gulf Coast of the southeastern United States, with peak production of approximately 12,000 tons of oil per year. These plantations were not competitive on the world market due to high labor costs, so that plantations have been abandoned. Tung-oil trees persist in these old sites. The tung-oil tree and the related candlenut tree (*A. moluccana*) are planted as shade trees in warm areas of North America.

Toxins: Tung-nut oil is composed primarily of oleostearic acid, which, beside being used as a waterproofing agent, has been known since ancient times in China as a powerful purgative. Saponins and phytotoxins have also been tentatively identified in the tung nuts, but not completely characterized.

Symptoms: As soon as 20 minutes after eating the pleasant-tasting seeds, the victim begins to experience nausea and then severe gastrointestinal irritation consisting of vomiting, diarrhea and stomach cramps. Severe diarrhea may lead to dehydration. Further symptoms include shock and kidney failure, and death in severe cases.

First Aid and Medical Treatment: Induce vomiting, replace fluids as necessary and monitor kidney function.

DOGBANE FAMILY (all parts) rarely ☠

Dogbane plants were formerly added to dog food to poison offending canines. Let us beware of these plants, lest we poison ourselves by accident! The dogbane family (Apocynaceae) is a large group of more than 1,000 species of primarily tropical trees, shrubs, herbs and vines. These plants may be easily recognized by the typical combination of simple, untoothed leaves produced in pairs, an abundant milky or at least gummy sap, petals forming a tube with a twisted look to the lobes and the fruit composed of one or two dry, brown cigar- or pencil-shaped containers filled with numerous small seeds, often with small tufts of hair. The dogbane family contains numerous highly poisonous species, such as oleander (*Nerium oleander*), yellow nightshade (*Urechites*), yellow oleander (*Thevetia peruviana*), and crape jasmine (*Ervatamia coronaria*). Even common species from which no cases of poisoning are listed, such as the dogbane genus *Apocynum*, should

84 Poisonous Plants and Mushrooms of North America

be treated with caution. The reported toxins from the family include cardiac glycosides and alkaloids. Oleander and yellow oleander are treated in greater detail.

OLEANDER (all parts dangerous) ☠

Oleander, *Nerium oleander,* is an attractive but deady Eurasian shrub or small tree that has been introduced throughout much of the American sunbelt, Hawaii, and the Caribbean for its attractive evergreen foliage and showy flowers. Oleander hedges have been used by landscapers in warm climates, and in the colder parts of the country it is grown as a house plant. The oleander's dark green, leathery leaves are long and narrow, with a short stalk and a prominent, light-colored mid-rib. The leaves are arranged in pairs along the stem or in whorls of three. Large, attractive, white, pink, or red blossoms are produced in large clusters at the end of the branches and each flower may give rise to one or two long, cigar-shaped seed pods. The sap is thick and clear, unlike that produced by most other members of the poisonous dogbane family. (**Plate 17.**)

Toxins: The deadly action of the oleander is due to a potent brew of over 50 toxic compounds, with the most important of these being two cardiac glycosides, oleandroside and nerioside. The action of these compounds is similar to the digitoxins found in foxglove (see p. 55). These poisons cause increased contractility of the heart muscle which eventually leads to overstimulation and loss of coordination of the heart muscle.

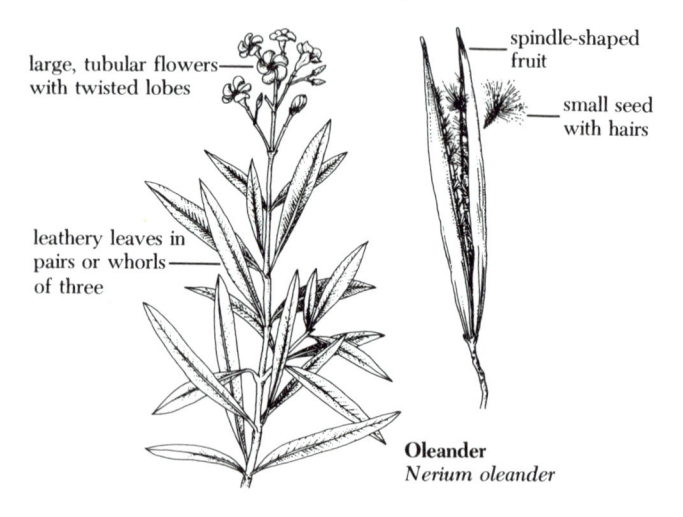

large, tubular flowers with twisted lobes

spindle-shaped fruit

small seed with hairs

leathery leaves in pairs or whorls of three

Oleander
Nerium oleander

Symptoms: The poison is contained in all plant parts, whether fresh or dried. Even eating a single leaf may be fatal. The poison is also present in the smoke coming from burning branch clippings and leaves and is even found in the honey made by bees visiting oleander flowers. Barbecuing meat with sticks made from oleander twigs has caused fatalities in unwary picnickers. Immediate toxic symptoms—burning in the mouth—are due to irritant saponins. Symptoms begin about one to two hours after ingestion and include nausea, vomiting, stomachache, dizziness and bloody diarrhea. One peculiar symptom noted in oleander poisoning is a visual perceptual change, the appearance of yellow and green colors and geometric patterns around objects. Further symptoms include drowsiness, loss of consciousness, slow and irregular pulse, difficulty in breathing and respiratory failure resulting in death.

First Aid and Medical Treatment: Treat for digitalis poisoning, see p. 6.

YELLOW OLEANDER, LUCKY NUT, BE-STILL TREE
(all parts, particularly the leaves)   dangerous ☠

This tropical American small tree or shrub *(Thevetia peruviana)* is commonly cultivated in Hawaii and subtropical North America for its dense green mass of very narrow, three- to six-inch-long leaves. The twig ends are crowned with large, yellow to orange trumpet-shaped flowers with five lobes. The yellow or red fruit,

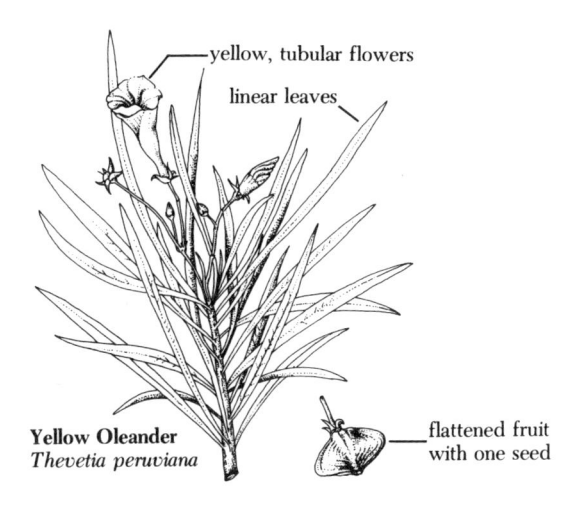

Yellow Oleander
Thevetia peruviana

yellow, tubular flowers
linear leaves
flattened fruit with one seed

which blackens at maturity, is flattened and nearly triangular, with a thin layer of flesh surrounding an inner bony part in which are found two large seeds. These seeds or "lucky nuts" may be sought after as good luck charms.

Toxins: The yellow oleander contains a cardiac glycoside, thevetin, similar in structure and activity to digitalis. There are reports of deaths from eating a single leaf.

Symptoms: See digitalis poisoning under foxglove. Yellow oleander is considered to be the most frequent cause of poisoning in Hawaii, often through eating "lucky nuts."

First Aid and Medical Treatment: See digitalis poisoning, p. 6.

RHODODENDRON FAMILY   (all parts) rarely ☠

The ornamental bushes surrounding our houses may be deadly strangers waiting to kill the unwary. Many of our common ornamental shrubs of cooler areas and related wild species are members of the heath family, the Ericaceae. Despite their diversity of flower colors and leaf types, many of these plants possess a deadly poison that was used by the Delaware Indians as a suicide potion. The poisonous shrubs of this family generally have dark evergreen leaves without teeth or with small teeth, bell-shaped flowers, and a small, dry, capsular fruit. The most common poisonous shrubs include species of the genus *Andromeda*, Labrador tea (*Ledum* sp.), Sierra laurel (*Leucothoe* sp.), and Japanese pieris (*Pieris japonica*), including the rhododendrons, azaleas and laurels treated below. The rhododendron family also includes many harmless and delightful edible fruit species such as blueberries, huckleberries and cranberries. You may eat these without worry.

In the spring, rhododendrons and azaleas almost appear to explode with dazzling clusters of flaming orange, yellow, pink, purple, or white flowers. The flowers vary in shape from that of a cup or tube to that of a funnel, depending on the type of pollinator that the plant is attracting. Each of these colorful flowers is unfortunately followed only by a small dull-brown capsular fruit. Following flowering, the shrubs present a cool, satisfying appearance with their generally large dark green leaves clustered at the end of long twigs. Species can be roughly divided into two groups: the rhododendrons, with smooth evergreen leaves, and the azaleas, with hairy deciduous leaves. (**Plate 19.**)

Distribution and Habitat: Numerous species of rhododendron and azalea occur in woods, swamps and mountain ridges, often forming thickets primarily in cooler and moister areas of the Ap-

87 Toxic Plants of the Home and Garden

palachian and west coast mountains. Many ornamental azaleas have been introduced from Asia.

Toxins: These species contain a white, amorphous or crystalline resin-like material called andromedotoxin, and a glycoside compound named arbutin. The andromedotoxin acts first by stimulating and then by blocking the nerves of the heart, leading to death by heart failure.

Symptoms: Poisoning may occur from making teas out of the leaves or twigs of these species or eating the bitter honey collected by bees from the flowers. Children may become poisoned by sucking nectar from the flowers. Initial symptoms begin with a temporary burning sensation in the mouth; after about one hour, victims may vomit and experience stomach cramps, diarrhea, excessive tear and saliva formation, a running nose and a piercing sensation in the skin. They may experience headache, general weakness, slow pulse (to 30 bpm or less), a severe drop in blood pressure, confusion, visual disturbances, shallow respiration, numbness and paralysis of the limbs, convulsions and rarely death. Paradoxically, some victims may develop dangerously high blood pressure.

First Aid and Medical Treatment: Induce vomiting and administer activated charcoal. Monitor EKG and blood pressure and be alert for marked hypotension. For hypotension administer approximately 2 mg of atropine subcutaneously and repeat if necessary until the heartbeat and blood pressure are normal. Be alert for paradoxical hypertension.

Individuals exposed to low doses through prolonged drinking of poison tea, honey or herbal medicines may develop chronic low pressure and episodic high blood pressure. Symptoms should disappear once the source of poison is discovered and eliminated.

LAMBKILL, SHEEP LAUREL, (all parts)
MOUNTAIN LAUREL, BOG LAUREL rarely ☠

The shrub laurels of the genus *Kalmia* are well known as killers of sheep, and can just as easily kill a person. The delicate pink to white flowers are produced in clusters so lovely that the mountain laurel has been named the state flower of Connecticut and Pennsylvania. The ten stamens are held under tension in little pockets on the bowl-shaped flower, and when a bee lands on the flower to collect nectar, each stamen catapults forward, showering the bee with pollen. The leaves are either alternate or opposite along the stem, and usually produced in clusters at the twig tips. The thick evergreen leaves are leathery and without marginal

showy, pink to white flowers

persistent capsules

evergreen, leathery leaves

Mountain Laurel
Kalmia latifolia

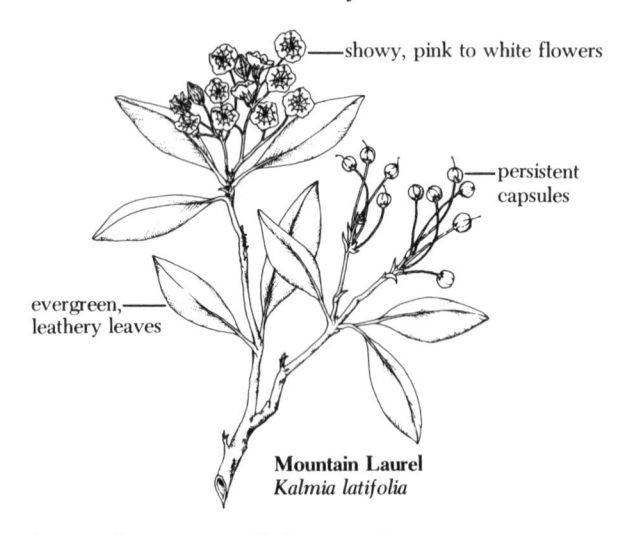

teeth. The fruit is a small five-parted capsule with many tiny seeds. (**Plate 18.**)

Distribution and Habitat: The mountain laurel *(K. latifolia)* is a medium- to large-sized shrub in moist rocky woods, in thickets, and on mountaintops throughout eastern North America. The mountain laurel is widely planted in gardens, particularly in front of houses. The sheep laurel or lambkill *(K. angustifolia)* and bog laurel *(K. polifolia)* are small shrubs to three feet tall with strongly ascending branches, which occur throughout North America in open habitats with poor soils, wet meadows, bogs and thickets.

Toxins and Symptoms: See rhododendrons (p. 87).

DAPHNE, MEZEREON, SPURGE LAUREL (fruits) dangerous ☠

This shrub *(Daphne mezereum)* receives its name from the lovely wood nymph Daphne, of Greek mythology who was changed to a small tree while fleeing from the lecherous Apollo. The name is quite appropriate for this plant with its hauntingly fragrant flowers which are blushing pink (rarely white) in color. In the spring, the flowers appear in little bouquets along the twigs before the lance-shaped, short-stalked leaves expand. The flowers are distinctively four-lobed, but examination reveals that the flowers lack petals, and the showy parts are really the sepals. The bright red or yellow, single-seeded fruits are produced on the twig below the terminal cluster of leaves.(**Plate 20.**)

Distribution and Habitat: Both this and other *Daphne* species, all originally from Europe and Asia, are frequently cultivated in North America, occasionally escaping to the wild.

Toxins: *Daphne* contains numerous poisonous compounds, including an irritating, blistering resin named mezereinic acid anhydride, as well as daphin, a bitter glycoside containing coumarin. Other extracts of *Daphne* have shown potential in treating cancer, in particular leukemia.

Symptoms: All parts contain poison, with the attractive fruits being particularly dangerous. Most accidents occur by children eating the fruits, with as few as ten fruits constituting a potentially lethal dose. Initial symptoms are a burning sensation in the mouth and throat, difficulty in swallowing, and swelling and inflammation in the eyelids and nasal passages, followed by vomiting, stomach pains, extreme thirst and diarrhea, often with blood. The patient may not be able to eat or drink. Kidney damage may develop, giving rise to blood or albumin in the urine or difficulty in urination. In severe cases, the victim may show delirium, headaches, convulsions and shock, leading to death. The sap of the plant may cause mild to severe skin irritations, almost approaching ulcerations. The poison may enter the body through skin contact with the poison, leading to the full range of symptoms.

First Aid and Medical Treatment: Induce vomiting and administer activated charcoal and cathartics. Later administer a de-

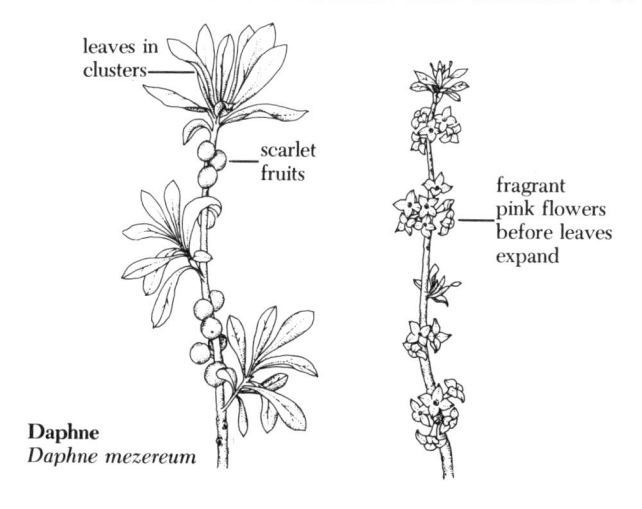

Daphne
Daphne mezereum

leaves in clusters

scarlet fruits

fragrant pink flowers before leaves expand

mulcent to soothe the irritated gastrointestinal tract. Electrolyte and fluid balances should be monitored carefully. The severe symptoms may last for several days, with an approximately 25 percent mortality rate for victims. This is clearly a plant to treat seriously!

CHERRY AND APPLE FAMILY (seeds, rarely)

Cherries, apples, plums and peaches are among our most important and delicious fruit trees and ornamentals, yet beneath the sweet and juicy, fleshy exterior of the fruit lies a seed that can kill. All of these members of the Rosaceae family have developed seeds that contain potent concentrations of cyanide, so that birds and man have learned just to eat the flesh and drop the seeds. While these trees are cultivated in gardens and orchards, and their fruit is well known, wild cherries are also important trees of our forests, particularly following fires or other damage to growing trees. The common black cherry *(Prunus serotina)* can be recognized by its shiny dark bark with horizontal corky slits, alternate lanceolate leaves with small marginal teeth and hairs along the midrib of the lower leaf surface and leaf stalks often with two reddish glands. The white five-petaled flowers are produced on an elongated stalk and develop into black fruit with a single large seed.

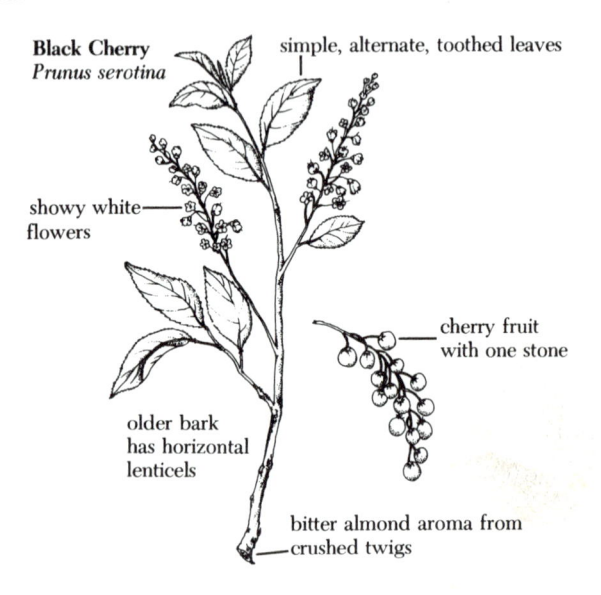

Black Cherry
Prunus serotina

simple, alternate, toothed leaves

showy white flowers

cherry fruit with one stone

older bark has horizontal lenticels

bitter almond aroma from crushed twigs

91 Toxic Plants of the Home and Garden

Distribution and Habitat: The cultivated species are grown throughout North America, with wild species also widely distributed.

Toxins: The bark, twigs and seeds of many members of this family contain the glucoside amygdalin which produces cyanide upon eating. The bark of black cherry is a frequent addition to herbal teas used as mild sedatives and general tonics. The value of these remedies is questionable and potentially dangerous. The controversial cancer medicine laetrile is made from extract of apricot pit, which has a high concentration of cyanide.

Symptoms: Symptoms develop after eating and chewing large numbers of seeds of apple, cherry or peach, or other members of the family. Almonds may be toxic if eaten in abnormally large quantities. Poisoning may also occur after eating or making a tea from the bark, leaves or twigs. Symptoms of cyanide poisoning can occur suddenly, and include difficulty with breathing, convulsions, general distress, gasping, coma, and death.

First Aid and Medical Treatment: See section on cyanide poisoning, p. 6.

PRIVET   (fruits) rarely ☠

One of the most widely planted hedges around houses and other buildings is the common European privet *(Ligustrum vulgare)* as well as related privet species. When well clipped, its dense mass of evergreen foliage make it an ideal screen. The lanceolate, glossy leaves are produced in pairs on short stalks, and many forms with variegated leaves are grown. The small, highly fragrant, tubular white flowers are produced in the summer in pyramidal clusters. The black, hard, popcorn-sized fruits remain on the hedges through the fall into the spring, making them available to children for all sorts of games.

Distribution and Habitat: Privet is widely planted in North America, escaping from cultivation in eastern North America.

Toxins: The bitter-tasting berries, bark, and leaves contain the glycoside ligustrin.

Symptoms: After eating the berries, the victim may experience vomiting and diarrhea, with convulsions and death in severe cases. Instances of privet poisoning are rare in North America.

hard, black berries

small white flowers

leaves in pairs

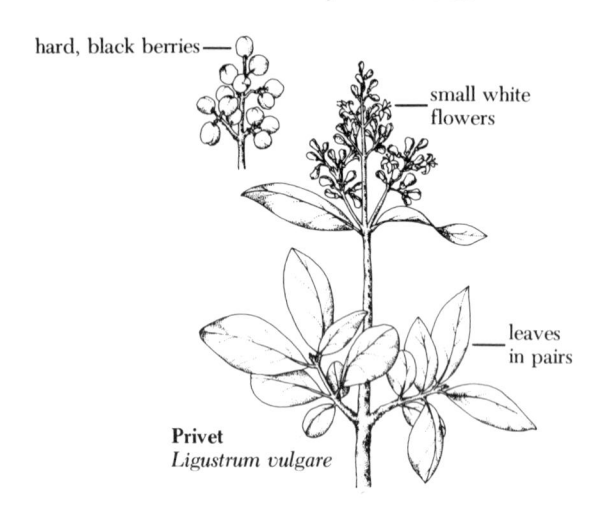

Privet
Ligustrum vulgare

GOLDEN CHAIN (seeds, flowers) rarely ☠

The golden chain, *Laburnum anagyroides*, a native of Europe, is cultivated as a small tree to 30 feet tall, primarily in the eastern United States, and is hardy into southern Canada. The long-stalked alternate leaves are divided into three leaflets which are slightly hairy on the lower surface. The sweetpea-shaped flowers are golden yellow and hang in pendant masses up to one foot long. The seeds are contained in long, flattened pods.

Toxins and Symptoms: Children may eat the attractive flowers or the pods, which may look to them like edible bean pods. However, the seeds and flowers contain cytisine, an alkaloid similar in its actions to nicotine. The poison contained in 20 seeds may be a fatal dose for a small child. Ingestion rapidly results in vomiting, diarrhea, drowsiness, weakness, poor coordination, sweating, pallor, headache, dilated pupils and a rapid heartbeat. Death may result from circulatory or respiratory failure. Golden chain poisonings are rarely serious because only small quantities of cytisine are usually ingested.

First Aid and Medical Treatment: See nicotine poisoning p. 112.

Cassava, Manioc, Tapioca, Yuca   (roots) rarely ☠

The cassava *(Manihot esculenta)* is a native of the tropical Americas, extending its range into south Florida and the Gulf Coast. It is widely cultivated throughout most of the tropics, particularly in Africa, where its large, elongated, sweet-potato-like tubers (up to three feet long and 10 inches in diameter) are a food staple. The up to nine-foot-tall, thick, above-ground stem makes this shrubby herb look somewhat similar to the castor bean plant. The long-stalked leaves are palmately veined, with about five to nine deep lobes. While rarely flowering, the flowers and the globose capsule are inconspicuous. All parts of the plant contain a milky juice.

To eat the tuber, the skin of the cassava roots must first be peeled off, and then the root itself must be boiled or baked to remove the poison. The cooked root yields a pleasantly bland, starchy vegetable that is similar to the potato. Tapioca is prepared by grinding up the root and pressing out the juice, followed by soaking and then boiling the remaining pulp until small balls of starch form. The cooked leaves of this plant are also eaten as a green vegetable.

Toxins: The roots contain significant quantities of cyanogenic glycosides, which are broken down in the body after eating to form cyanide. Proper preparation involving heating renders the cyanide inactive. However, young children who are unfamiliar with cassava may attempt to eat uncooked or inadequately cooked cassava roots and leaves and be poisoned. Cyanide is a potent toxin that interferes with cellular respiration and oxygen utilization. This great toxicity was dramatically demonstrated when the followers of the Reverend Jones used cyanide to commit mass suicide. Even prolonged exposure to very low levels of cyanide can have detrimental effects. A diet composed almost exclusively of cassava is found in some areas of Africa and leads to an increased incidence of diabetes through the selective effect of the cyanide on the pancreatic islet cells.

Symptoms: After eating a piece of raw or inadequately cooked root, the victim may begin to breathe more rapidly and become excited or even start gasping for breath in severe cases. Further severe symptoms may include physical collapse, convulsions, bluing of the lips and fingernails and coma, with death occurring within a few hours.

First Aid and Medical Treatment: See cyanide poisoning p. 6.

SPURGE FAMILY

RED SPURGE (tissue irritant)

Members of the Euphorbiaceae can assume the form of a weed, shrub, tree or succulent, but they all usually have one thing in common: their highly irritating milky sap. *Euphorbia cotinifolia*, the red spurge, is a small to medium shrub (up to 20 feet tall) with red branches covered with colorful long-stalked leaves, the upper portion of its leaves often being purple or maroon. When it gets dry or cold, the plant drops its leaves and looks rather bedraggled. They do occasionally produce small white flowers and three-lobed, hairy fruit. The plant's caustic sap produces inflammation of the skin and moist membranes, even causing temporary blindness if it gets in the eyes. Some people have become sensitized to it, and for them even brushing against the leaves can cause inflammation or blistering of the skin.

CROWN-OF-THORNS (mechanical injury, irritant sap)

A common inhabitant of homes, the crown of thorns, *Euphorbia milii*, is a decorative but fierce-looking ornamental plant. Its many brown to blackish branches are covered with long, very sharp spines. Its leaves are oval, and each branch is frequently tipped with a rosette of small flowers surrounded by two lip-shaped, bright red bracts. The plant can produce painful puncture wounds and its milky sap is a potent irritant of skin and moist membranes.(**Plate 21.**)

PENCIL TREE CACTUS OR MALABAR TREE (tissue irritant)

The pencil tree, *Euphorbia tirucallii*, is a succulent, cactus-like tree that forms a mass of slender, outward-curving, green, rubbery branches, often tipped by a few tiny, slender, lance-shaped leaves. The branches are pencil-thin; hence the name. In the north, it is grown as a household ornamental and in the south as an outdoor ornamental that reaches heights of thirty feet and spreads of twenty feet. Although the plant flowers and bears fruit in its native habitat, Africa, in gardens in the southern United States and used as a house plant, this doesn't happen. It grows easily from cuttings, and because of its rapid growth will require pruning, and herein lies the problem, for its white, milky thick latex sap contains some very potent, caustic irritants that can cause dermatitis. If the sap

95 Toxic Plants of the Home and Garden

Spurges

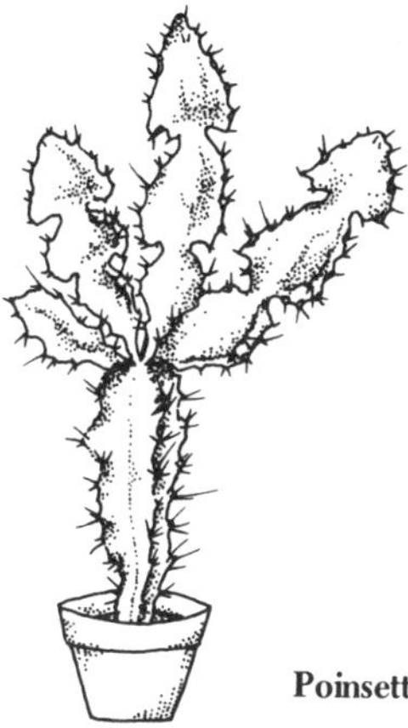

Spurge Cactus

Euphorbia frankiana

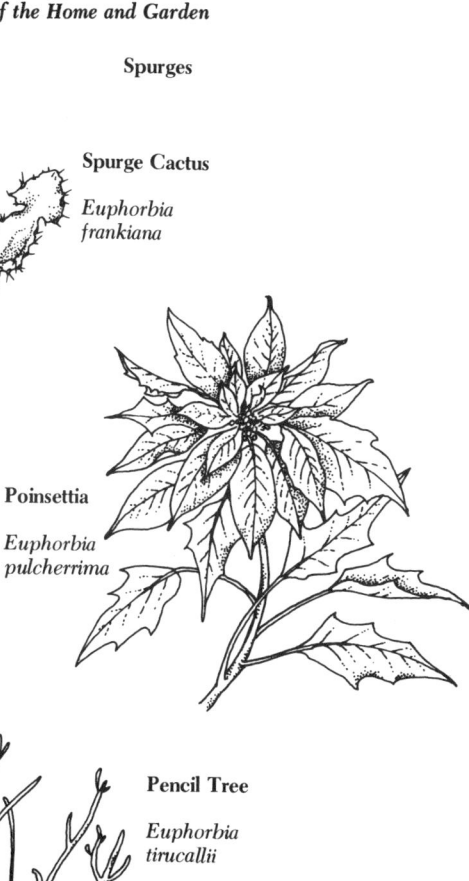

Poinsettia

Euphorbia pulcherrima

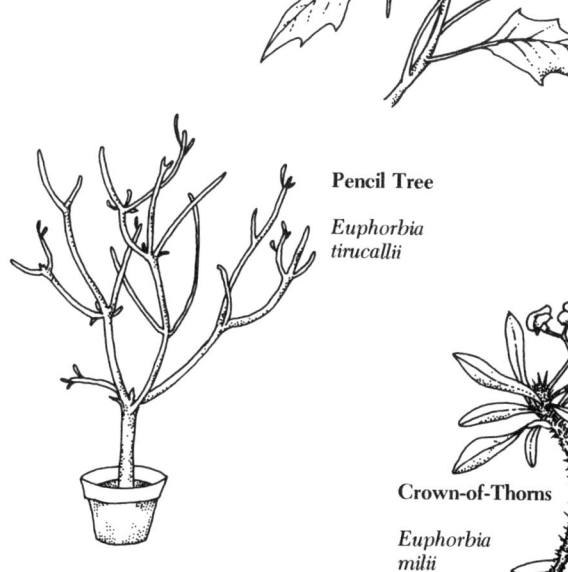

Pencil Tree

Euphorbia tirucallii

Crown-of-Thorns

Euphorbia milii

gets into the eyes, it causes painful, acute conjunctivitis, and if put in the mouth may cause extreme irritation of the lining of the mouth and tongue. A few cases of ingestion by children have been reported in Florida.

CANDELABRA CACTUS
(mechanical injury, irritant dermatitis)

Another succulent member of the family Euphorbiaceae is the candelabra cactus, which is not a real cactus but *Euphorbia lactea*. The tall, erect, tree-like plant has a straight trunk which sends out three or four angled branches, giving it the appearance of a candelabra. It may reach heights of 15 feet. Some varieties have crested or fluted branches and are much in demand as indoor ornamentals. This East Indian native thrives in warm climates and is found from southern Florida through the Caribbean. It has very sharp spines protruding from each crest, and brushing against or leaning into one of these plants is a very unpleasant experience. Indeed, they have been used to form protective barrier hedges around private homes because of their potential for creating mechanical injury. If the branches are broken, they exude a milky, caustic, irritant sap which damages skin and moist mucous membranes on contact.

SNOW-ON-THE-MOUNTAIN (tissue irritant)

Another of the spurge family with an acrid irritating sap is "snow-on-the-mountain," *Euphorbia marginata*, a native of the western prairies of the United States. This plant has become widely used as an ornamental because of its handsome white-margined green leaves and showy white petal-like bracts (actually modified leaves). Snow-on-the-mountain is a bushy, highly branched, two-foot-tall herb. Its oblong stalkless leaves (about two inches long) have white margins. The flowers are inconspicuous and tiny; the clusters of white bracts give the plant its name. The plant develops a small, round, three-lobed capsular fruit.

Distribution and Habitat: The plant is native in the prairies from South Dakota southward to Texas. It is used as an ornamental in the east, but has escaped cultivation and is found in the wild in warmer climates.

Toxins and Symptoms: The acidic, irritant, white latex sap can blister the skin, irritate cuts and cause pain. It is also very irritating to the eyes and other moist membranes, the mouth, and nasal passages. Eating honey produced by bees that feed on this or any

97 Toxic Plants of the Home and Garden

of the spurges can cause irritation of the bowel and typical G.I. symptoms.

POINSETTIA (tissue irritant)

Handsome Christmastime inhabitants of northern homes and full-time outdoor residents of warmer climes are the several varieties of poinsettia, *Euphorbia pulcherrima*. The common and most popular variety is prized for its brilliant red bracts (actually leaves) at the top of the stem, with the actual flowers being little green and yellow nubs in the center of the cluster of bracts. Some varieties have white or salmon-colored bracts. As is the case with all Euphorbiaceae, its white milky sap is irritating, but the poinsettias, like many a handsome specimen, has a very bad reputation which may be undeserved. In 1919 there was the report of a child in Hawaii dying after eating poinsettia leaves, and in 1965 there was a two-and-a-half-year-old in Rochester, New York, who became ill after eating poinsettias. However, the leaves are bitter tasting, and most children will spit them out. They will irritate the mouth, and the sap is a potent irritant of the eyes. Some laboratory studies with extracts of the plants failed to show any highly toxic substances, but any potent irritant will produce gastrointestinal symptoms if swallowed. Still, to be on the safe side, keep the plant out of the reach of small children.

OTHER ANNOYING SPURGES

Other common weeds of this genus that exude irritant latex are the flowering spurge, the cypress spurge and the caper spurge (*E. lathyris*), whose fruits have been mistaken for capers. Even though they have been pickled, the toxin is unchanged, and ingesting them can make the ingester very sick.

General Treatment for Spurge Toxins: Wash latex off as soon as possible and treat irritation symptomatically. If ingested and the victim has not vomited, induce vomiting and administer activated charcoal. Otherwise treat symptoms as they occur.

PART IV
Wild Poisonous Plants

EACH year, poisonous wild plants are responsible for the loss of tens of millions of dollars to commercial growers of range animals—such as cattle, sheep, goats and horses. Wild plants are also responsible for large numbers of human cases of gastrointestinal upset that range from mild to severe. Some plants contain very deadly poisons that have caused serious illness and death. Three populations are most seriously at risk. First, of course, are young children, whose inordinate curiosity, behavior and small body mass make them primary targets. Many of these poisonous plants produce attractive fruits and flowers that pack a potent punch, and young children have an unfortunate penchant to put a great variety of things into their mouths. Fortunately, most of these toxins normally produce only transient symptoms which usually persist for only a few hours to a day or two. However, some plants are very dangerous. If your child eats a plant which you suspect is poisonous, you should contact your local poison center and bring in a sample of the plant.

The second high risk group is made up of those who revel in natural edibles which add variety, flavor and nutrients to their diets. Among the edibles are plants used in making soups and purées, starchy root vegetables and nuts, cooked greens, salads, condiments, syrups and jellies. Unfortunately, plant identification is not stressed in many books on edible plants, and many an edible wild plant has some rather deadly look-alikes; an error in identity could be unpleasant or even lethal. Unfortunately, some gurus of wild foods have written guides that not only fail to warn you of poisonous look-alikes but even recommend some plants that are frankly poisonous. Anyone who wishes to fully enjoy the gourmet delicacies of nature should also know the risks and be able to identify with certainty both the poisonous and the edible plants, so that the risk-benefit decision favors the eater. In this section is information about all of the major dangerous wild plants and most of the less dangerous ones.

The third group of outdoors people who are at risk from wild plants are those who use natural herbals as medicines. The use of herbal medicines has a long history, and many folk medicines have been incorporated into modern medical pharmacology. Unfortunately, some of the advocates of herbal medicines who have written books on the subject are either irresponsible or terribly misinformed, because some of them have enthusiastically recommended supposedly curative plant products that are very dangerous—for example, May apple juice. Users of natural or herbal medicines have three problems: The first is accurate identification of the plant in question—unfortunately, some rather innocent and even beneficial plants have poisonous cousins who mimic their appearance. Second, some of the reputed beneficial plant medicinals are either questionably safe or even known to be dangerous. Third, plants may vary greatly in concentration of actively poisonous ingredients, due to the soil, local climate, plant age, and genetic type of the plant in question. As a result, dosage of medicine being administered cannot be known with certainty, and the patient receiving a certain weight or volume of herb medicine may be receiving an inadequate dose or an excessive dose. Such considerations are important, since many drugs reach their point of medical usefulness at levels only slightly below levels which are dangerous. This guide will provide the careful herbalist with a very accurate description of those plants which could poison instead of cure. If you feel you must use your own herb medicines, particularly if they are listed in this book, do so under the supervision of a trained physician, who can determine if you are contemplating an overdose.

CELANDINE POPPY, ROCK POPPY (sap, all parts) rarely ☠

The rock poppy, *Chelidonium majus*, a widespread Eurasian weed, is a perennial plant with deeply lobed green leaves that are pale on their undersides. During much of the year the plant appears as a rosette of leaves near to the ground, but during the spring, this rosette sprouts a tall branched stem (up to three feet high) with numerous bright yellow flowers, each of which has four petals and many stamens. These eventually give rise to clusters of upright green pods each about two inches long. All plant parts contain a bright orange or yellow milky sap, but the greatest concentration of this sap is in the root. The plant derives its scientific name from the Greek word for the swallow, *chelidon*, because the ancient scholars believed that mother swallows washed the eyes of their babies with the yellow juice of this plant to make their eyes strong.

100 Poisonous Plants and Mushrooms of North America

- finger-like clusters of fruit
- yellow flowers with four petals
- leaves deeply lobed or divided
- clasping leaf stalk
- orange sap in all plant parts

Celandine Poppy
Chelidonium majus

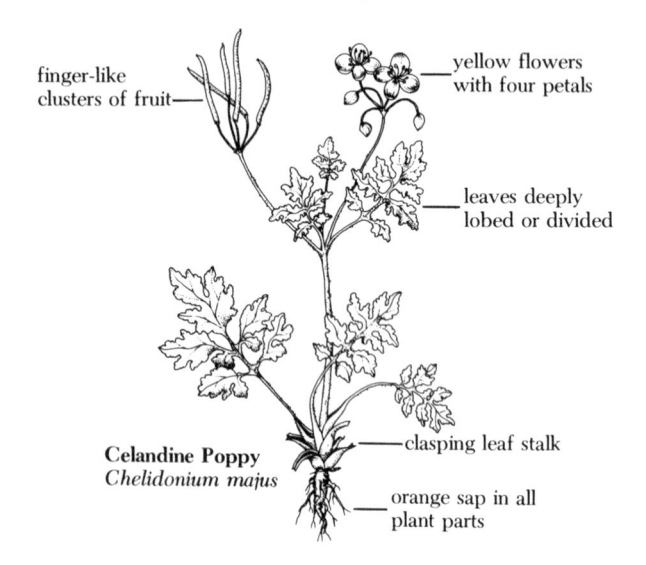

Distribution and Habitat: The rock poppy is found throughout eastern North America in disturbed areas usually with some shade, such as woodland paths or field edges, and near buildings.

Toxins and Symptoms: The sap is a potent irritant of the skin and moist membranes, and is capable of causing painful inflammation of the eyes. If ingested, the plant causes an almost immediate irritation of the mouth, throat and digestive tract causing pain, vomiting, and dizziness. As the toxin is absorbed there may be generalized edema and diarrhea. Emptying the bladder produces a burning sensation, and there may be blood in the urine. In severe cases, circulatory failure may lead to coma and death. The toxins present in this plant have been reported to damage the liver, lungs and bladder, and somehow affect the nervous system. Historically, the celandine poppy had been used in European folk medicine to treat skin growths such as warts and corns. The celandine poppy toxins are a mixture of alkaloids whose mode of action is not fully understood, but some of these compounds are currently undergoing testing in several countries for their potential as cancer-fighting drugs.

First Aid and Medical Treatment: If the patient has not vomited, induce vomiting and administer activated charcoal. Where dehydration has occurred, fluids, salts and glucose should be administered parenterally. Urine output should be monitored. Topical cortisone creams can be used to treat skin inflammation.

101 Wild Poisonous Plants

Related species: Other wild members of the poppy family are the prickly poppy *(Argemone mexicana),* found as a weed from tropical America into the southern United States, and the bloodroot *(Sanguinaria canadensis),* found in rich woods in eastern North America. Both these species have lobed leaves and a yellow or red-colored sap. Each can produce symptoms similar to that produced by the celandine poppy. If a plant oozes yellow, orange or red sap, watch out!

POISON HEMLOCK, WATER HEMLOCK, FOOL'S PARSLEY
(root stock, all parts) dangerous ☠

The flavorful family of such aromatic plants as carrot, celery, parsnip and parsley also gives us three species of our most deadly plants, which kill through imitating the appearance of our delicious vegetables: (1) Poison hemlock, also known as fool's parsley or California fern *(Conium maculatum).* (2) Water hemlock, spotted cowbane *(Cicuta maculata* and other *Cicuta* species). The names of these two species of poison herbs should not be confused with the common forest hemlock tree, *Tsuga.* (3) Fool's parsley *(Aethusa cynapium).*

The most famous victim of poison hemlock was the Greek philosopher Socrates, who was executed in 399 B.C. for teaching youths how to form their own opinions. Socrates said, "An unexamined life is not worth living," to which we could reply, "An unexamined plant is not worth eating." Dangerous members of

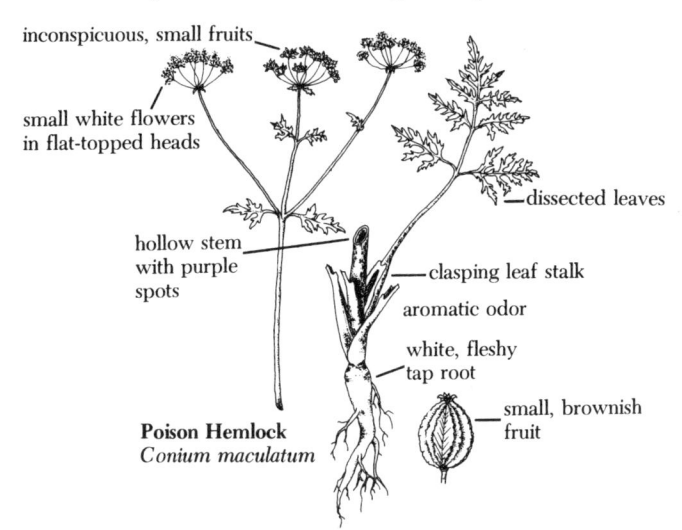

inconspicuous, small fruits

small white flowers in flat-topped heads

dissected leaves

hollow stem with purple spots

clasping leaf stalk

aromatic odor

white, fleshy tap root

small, brownish fruit

Poison Hemlock
Conium maculatum

102 Poisonous Plants and Mushrooms of North America

small white flowers in flat-topped clusters

dissected leaves

aromatic odor

hollow stem

clasping leaf stalk

Water Hemlock, Cowbane
Cicuta maculata

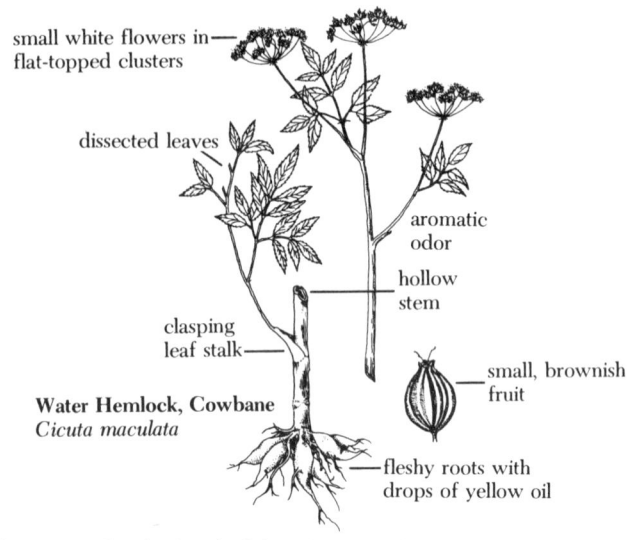

small, brownish fruit

fleshy roots with drops of yellow oil

the carrot family (Umbelliferae) are easily recognized by a few key distinguishing features.

These three fatal herbs all have an erect, hairless, hollow two- to-six-foot-tall stem with highly divided fern-like leaves. The stem may be slightly grooved, and, in the poison hemlock, is often spotted with purple. (**Plate 22.**) The leaf stalk clasps the stem as a celery stalk clasps the rest of the celery bunch. The crushed leaves and stalk usually have either an aromatic smell, like parsnip or parsley, or an unpleasant one. The flowers are borne in flat-topped white clusters. The fruits are small, inconspicuous, dry and brown with streaks or ribs along the length of the fruit. In fool's parsley, small leaf-like appendages like tiny green beards hang from the flower clusters. The biennial roots of the poison hemlock look like small white carrots. The roots and shoots of the water hemlock may produce a yellowish oil when cut. The perennial roots of the water hemlock consist of a bundle of several fat white tubers, in contrast with the single root of the parsnip. There are many edible cultivated and wild plants in this family as well, such as caraway, anise, wild carrot or Queen Anne's lace, cow-parsnip, parsley and angelica. Unlike the poisonous plants, wild carrots have hairy stems, and parsley has a yellow-green flower. But unless you are an experienced herb collector, don't gamble that you can distinguish these edible plants from the poisonous plants: if a plant has divided leaves, a hollow stem, clasping leaf bases, and a flat-topped cluster of white flowers, leave it alone.

103 Wild Poisonous Plants

Distribution and Habitat: Poison hemlock and fool's parsley have been introduced from Europe into North America, where they have become widely naturalized. Poison hemlock is found as a weed throughout North America, most commonly in the northern United States and adjoining Canada, in generally wet, disturbed ground such as ditches and the edges of fields. Fool's parsley is found as a weed in disturbed ground and gardens in northeastern North America. Water hemlock grows in damp open habitats, such as ditches, wet meadows and stream edges throughout eastern North America. Other native *Cicuta* species are distributed throughout North America.

Symptoms: Cases of poisoning occur through people mistaking these plants for edible plants such as parsley, celery, wild carrot, or parsnip and eating the roots, stems, leaves, or seeds. Children may be poisoned through using the hollow stems as whistles or pea-shooters.

From a few minutes to a few hours after eating, the victim experiences a burning sensation in the mouth, difficulty in swallowing, and foaming at the mouth. Internal symptoms then develop, such as vomiting, nausea, stomachache and diarrhea. As the poison begins to affect the nervous system, the victim may experience nervousness, muscle twitches, and sometimes violent clenching of the teeth, slowing of the heart, weak pulse, visual disturbances and gradual heart failure, leading to a loss of consciousness and death. Death may occur as soon as 15 minutes

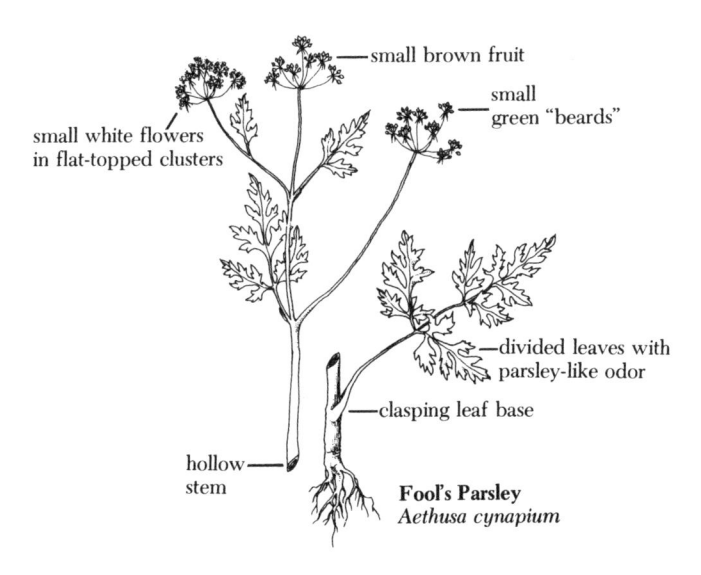

Fool's Parsley
Aethusa cynapium

after exposure or as late as eight hours after exposure. Cases of death due to eating fool's parsley are relatively rare. Following this serious period, patients may still experience muscle soreness and damage, kidney damage, and impairment of mental faculties.

Toxins: The toxic qualities of poison hemlock and fool's parsley are related to the presence in all plant parts of a group of nicotine-like alkaloids, the most important of which is coniine, a colorless, volatile oil. Coniine produces a temporary stimulation of the nervous system, followed by a general depression of the nervous system. Coniine appears to act by blocking nerve endings. Death is often quite painful because the victim remains conscious during the violent convulsions. The poison of water hemlock has been found to be an unsaturated alcohol named cicutoxin. The action of the cicutoxin is not well understood, but may involve an overstimulation of the nervous system, leading to muscular convulsions.

First Aid and Medical Treatment: If the subject is alert, induce vomiting and follow with activated charcoal. This may be difficult if there is clenching of the jaws and gastric lavage may be needed. Administer cathartics to clear poisons from the gastrointestinal tract. Barbiturates or Diazepam should be used to control the convulsions. Keep the air passage clear and be prepared to treat for respiratory failure. Sodium bicarbonate may be used to control metabolic acidosis. Hemodialysis may be necessary. Saline and colloids may be required to treat a drop in blood pressure.

POKEWEED, POKEBERRY, PIGEONBERRY (all parts) rarely ☠

Pokeweed *(Phytolacca americana)* is a large perennial herb which derives its common name from the American Indians, who called it *pocan* and used its juices for making red dyes. In the past, extracts from this plant were used as a medicine to induce vomiting and as a powerful laxative. In American folk medicine it was also used to treat some skin diseases, but the toxins contained in the plant were so powerful that its medical benefits were questionable.

Pokeweed has a large, fleshy taproot that in the spring sends up a cluster of foot-long, oblong, green succulent leaves that look as tempting as wild spinach or Swiss chard. From this cluster of leaves grows a thick, reddish, leafy stem up to nine feet in height. Later in the summer, the plant produces dropping, curved spikes of greenish-white flowers, which develop into dark purple berries, slightly larger than blueberries, that often attract unwary children who may eat them or crush them to extract their purple juices

Pokeweed
Phytolacca americana

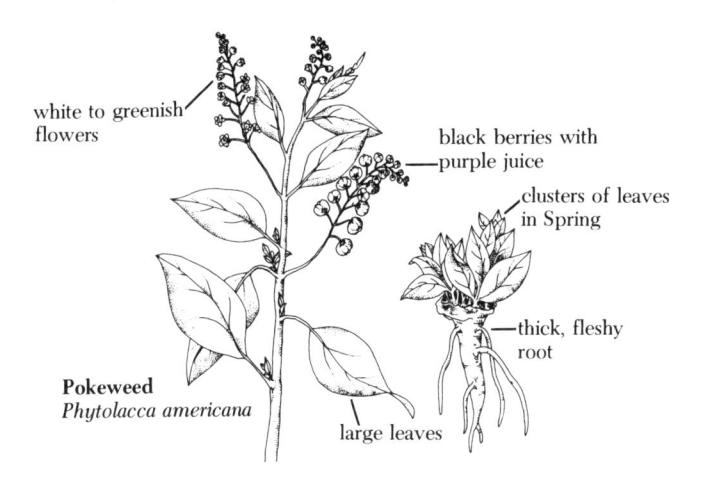

- white to greenish flowers
- black berries with purple juice
- clusters of leaves in Spring
- thick, fleshy root
- large leaves

to make a beverage that looks like grape juice. Children are the main victims of this handsome toxic plant, which as kids we used to call "the barfberry bush." (**Plate 23.**)

Distribution and Habitat: This native American weed is found commonly throughout eastern North America and occasionally in other parts of North America and even Hawaii, frequenting fields, roadsides, and disturbed ground. It is often found on rich soil around buildings and barnyards. A related poisonous species is *P. rigida*, which is found on sand dunes and marshes on the coast of the southeastern United States. There are also several native Hawaiian species.

Toxins: The major poisonous principle appears to be the saponin, phytolaccine, with other minor poisons also identified. Other proteins in pokeweed have been found to stimulate cell division of human lymphocytes in cell culture. This property is being actively studied by cancer scientists as a model system to study abnormal tissue growth. It is not clear if this type of cell-stimulating activity is related to the poisonous properties of pokeweed.

Symptoms: All parts of the plant are poisonous, but the roots seem to be the most toxic, and the berries the least toxic. Cooking the leaves as a vegetable in at least two waters and boiling the berries for making pies seems to render the poison inactive. However, eating raw or inadequately cooked pokeweed, in particular the root, can lead to disastrous results. The poison initially causes only a burning sensation in the mouth; however, after about two hours, the victims may experience severe stomach cramps, vomiting and diarrhea. In severe cases, symptoms may include difficulty in breathing, profuse sweating, salivation, dizziness, weakness, muscular stiffness, a weak pulse, visual difficulties and even convulsions. Most victims will recover within two days, although there have been a few cases that were fatal.

First Aid and Medical Treatment: No treatment has been found effective other than emptying the stomach contents.

DEATH CAMAS, BLACK SNAKEROOT   (bulbs, leaves)
rarely

The name of this group of plants is a clear tip-off that this is a plant to avoid if you are looking forward to collecting your full retirement benefits. These species are perennial herbs, with a dark onion-like bulb or a thick, creeping root-stock. Death camas (*Zigadenus* spp.) lacks any onion aroma in their leaves or bulbs, which distinguishes it from wild onions. Most of the one-foot-long, thin, slightly folded, grass-like leaves are produced in a

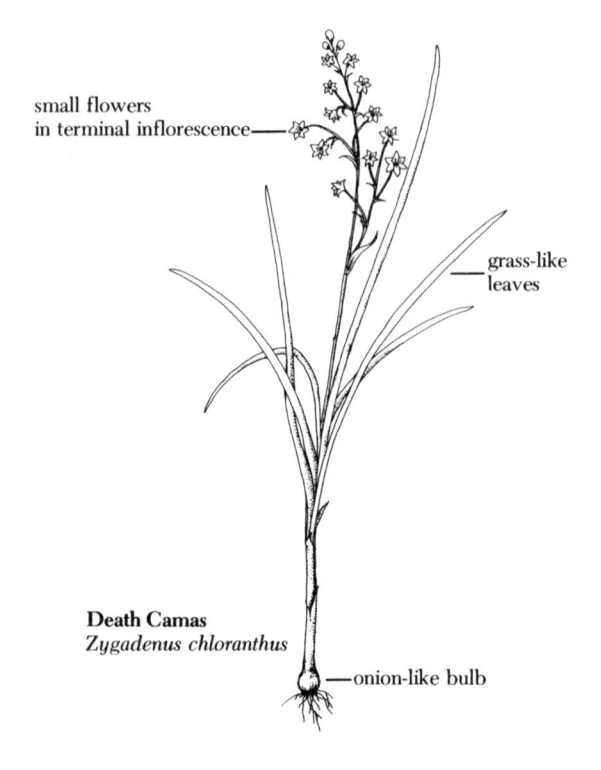

small flowers in terminal inflorescence

grass-like leaves

Death Camas
Zygadenus chloranthus

—onion-like bulb

bunch at ground level. A few smaller leaves clasp the unbranched flowering stalk, which may reach three feet in height. While the pink, white, greenish, or yellow flower cluster itself makes a good show, the individual flowers are small and develop into dull-colored capsules.

Distribution and Habitat: There are about 18 species of death camas known from North America and northern Asia, with exact species identification being difficult. Certain species appear to have little poison, so that after accidently eating a death camas and then saying your final goodbyes to friends and family, you may not show any symptoms. Death camas species occur in dry prairies, grassy hillsides, meadows, damp pine woods, and bogs.

Toxins: Death camas contain a mixture of steroid alkaloids, including zygacine. The death camas plants are responsible for most of the cases of poisoning of grazing livestock in the rangelands of western North America. In these areas, a special "Death Camas medicine" has been developed for poisoned sheep, consisting of 2 mg of atropine sulfate and 8 mg of picrotoxin per 100-pound sheep, given by injection or tablet.

Symptoms: Poisoning occurs through eating the bulbs or other plant parts. The victim experiences vomiting, stomach cramps, diarrhea and gradual weakening of the respiratory and muscular system, leading to respiratory failure.

First Aid and Medical Treatment: If the victim has not vomited, induce vomiting and administer activated charcoal. Monitor vital signs and replace fluids and electrolytes. Otherwise, treat symptomatically.

MILKWEEDS (*Asclepias* sp.)

Many different species of milkweeds are found widely distributed throughout North America, primarily in fields, roadsides, dry plains, swamps and riversides. The species are all perennial herbs with upright stems usually two to three feet tall, flowering in the midsummer to fall. The key identifying characters of milkweeds are a milky white sap coming from broken leaves and stems, and the arrangement of the leaves in pairs or whorls along the stem. The untoothed leaves vary in shape from linear to oval. (**Plate 24.**)

The flowers are produced in bunches from above individual stem leaves or at the top of the stem. The flowers may be bright orange, pink, white, or green. Each flower has a complex five-part construction in which the petals, stamens and pistil are fused together; at the top of the flower are five horns which fill with

Butterfly Milkweed
Asclepias tuberosa

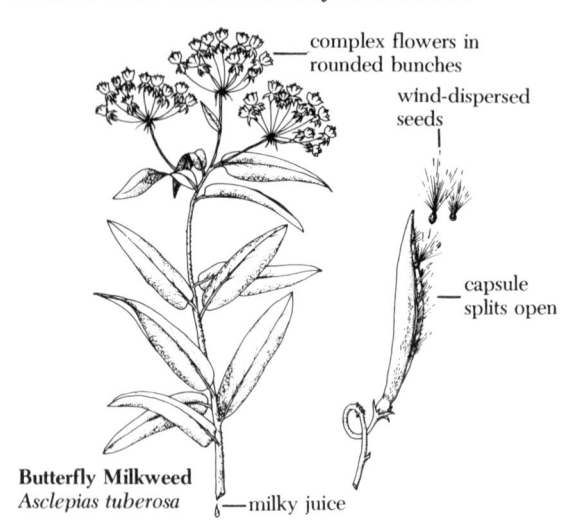

- complex flowers in rounded bunches
- wind-dispersed seeds
- capsule splits open
- milky juice

nectar. The fruit is a pod which splits open in the late fall, exposing the individual flat, pale brown seeds, which are wafted away on their silken fiber parachutes. The common milkweed, *A. syriaca*, which has pink or greenish-white flowers, has edible young stems, flowers, and pods. The butterfly weed, *A. tuberosa*, with brilliant orange, red or yellow flowers and a watery sap, is probably the most poisonous species of the milkweeds.

Toxins: Milkweeds have a mixture of poisons, the most important being a resin-like compound, galitoxin, as well as glucosides and alkaloids.

Symptoms: The patient usually complains of stomachache and shows symptoms of intestinal inflammation. In severe cases, fever and difficulty in respiration may develop.

Related species: Other poisonous plants, such as the dogbanes (Apocynaceae) and the spurges (Euphorbiaceae), also have milky juice. If a plant has milky juice, avoid touching or eating the sap!

First Aid and Medical Treatment: If the patient hasn't vomited, induce vomiting and administer activated charcoal followed by laxatives.

BUTTERCUP, CROWFOOT   (irritant sap) rarely ☠

Buttercups of the genus *Ranunculus* are very common, attractive, perennial or annual herbs of the large, widely distributed family

Ranunculaceae. Despite their innocent-sounding names and harmless appearance, they pack a potent toxin responsible for numerous poisonings, particularly among children. They are easily identified by their conspicuous bright yellow flowers with usually five (though up to seven) shiny petals and numerous stamens. Each flower can vary in size from an eighth of an inch to one inch in diameter, and develops into a fruit that is composed of a round cluster of green seeds. The plant itself has a cluster of leaves (palmately veined, with teeth and deeply cut lobes) at ground level. At flowering, a stem of up to three feet tall is produced, with alternate leaves and one or more flowers at the end of the stem (see cover, lower left).

Distribution and Habitat: Buttercup species are common and widespread throughout most of North America, but they are primarily found in wet meadows, streamsides, waste areas, and in cooler climates. Among the most commonly encountered species are the tall field buttercup *(R. acris)*, the bulbous buttercup *(R. bulbosus)*, the small-flowered buttercup *(R. abortivus)* and the creeping buttercup *(R. repens)*. The creeping buttercup is often locally abundant, forming large patches along the coastal areas of the Pacific Northwest.

Toxins: The poisonous substance found in buttercups is an acrid yellow oil, protoanemonin. Despite some natural food guides which suggest the protoanemonin is inactivated by heat, this poison may not be completely destroyed by cooking. Fortunately,

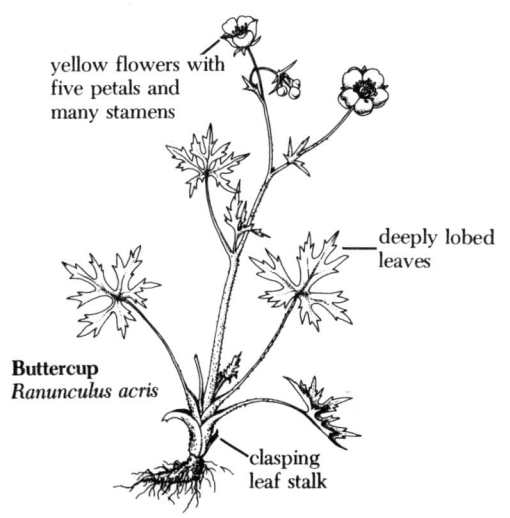

Buttercup
Ranunculus acris

- yellow flowers with five petals and many stamens
- deeply lobed leaves
- clasping leaf stalk

the poison is not highly toxic, and its concentration in plant parts is quite variable. Children are particularly attracted to this handsome yellow flower and play a game seeing if its reflection under the chin will foretell true love.

Symptoms: The sap of the plant contains variable amounts of an irritant poison that can cause blistering of the skin when the crushed plant parts are handled. Eating the plant may result in a burning sensation, followed by slavering, irritation, redness, and blistering of the mouth and throat, and, if swallowed, will cause vomiting, stomach pains and diarrhea. If enough of the toxin is absorbed into the body, generalized symptoms may develop that include dizziness, fainting, and, in rare cases, convulsions. The buttercup's poison may also cause kidney damage resulting, initially, in abundant urine production (polyuria), followed by blood in the urine (hematuria) and then limited urine flow. Death due to kidney failure may occur, but rarely does.

First Aid and Medical Treatment: Induce vomiting and follow by administering activated charcoal. Monitor renal functions and if needed give parenteral fluid and electrolytes.

Related Species: Several other members of the buttercup family (Ranunculaceae) also contain more or less quantities of protoanemonin and may give essentially the same symptoms. They include the following:

BANEBERRY, DOLL'S EYES, CORALBERRY, WHITE COHOSH

Species of *Actaea* are perennial herbs with large, highly divided leaves and red or white fruits on long, thin stalks. These plants occur in moist, rich woods throughout North America.

ANEMONE, WIND FLOWER, PASQUEFLOWER, THIMBLEWEED

Species of *Anemone* are small woodland and montane herbs with a single white flower on each stem.

MARSH MARIGOLD, COWSLIP

Species of yellow- and white-flowered *Caltha* look very similar to buttercups, and are widely distributed in North America in swamps, on stream edges, and in mountain bogs.

Virgin's-Bower

The attractive climbing vines of *Clematis* include both native and ornamental species which are suspected of being poisonous.

Christmas Rose

The introduced perennial *Helleborus niger* is grown as an ornamental for its attractive white flowers which bloom in winter. The plant contains the digitalis-like glycoside, helleborein.

Indian Tobacco   (leaves, stems) rarely ☠

Indian tobacco *(Lobelia inflata)* was dried and smoked by the American Indians, presumably as a hallucinogenic plant. Since then, Indian tobacco has been regarded as an important American medicinal plant for hundreds of years, based on the idea that the American Indians knew the true secrets of herbal medicine. Indian tobacco has been thought to be effective in treating asthma and bronchitis, as an emetic and antispasmodic and as an additive to anti-smoking pills. Indian tobacco is still a commonly added component of herb medicines, despite documented cases of deaths and harmful effects of its use. Indian tobacco is a small, erect, annual herb with alternate, simple, toothed ovate leaves. The small tubular flowers are usually blue, with long, thin, green sepals. The flowers are bilaterally symmetrical, with two lobes

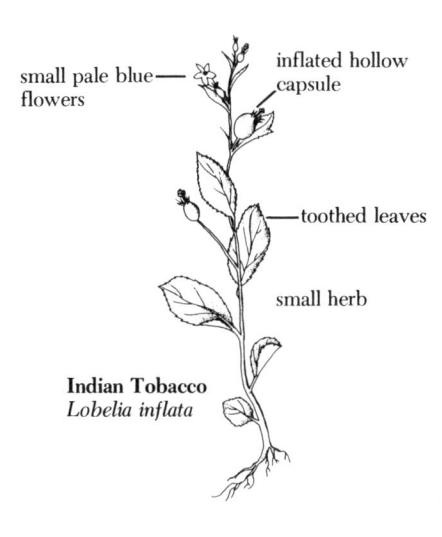

Indian Tobacco
Lobelia inflata

- small pale blue flowers
- inflated hollow capsule
- toothed leaves
- small herb

curving upward and three lobes curving downward. The fruit is an inflated capsule with small seeds.

Distribution and Habitat: Indian tobacco is a common herb of trailsides, forest clearings and fields in eastern North America. Related species of the genus *Lobelia* occur throughout North America. The lovely cardinal flower, *L. cardinalis*, with its electric red, hummingbird-pollinated flowers, grows in wet open ground in eastern North America and is often cultivated.

Toxins: Mixtures of pyridine alkaloids chemically similar to nicotine are found in all plant parts.

Symptoms: Poisoning usually occurs from individuals using locally made herb medicines containing excessive amounts of Indian tobacco. Initial symptoms may include nausea, vomiting, a general feeling of weakness and prostration. In severe cases the pulse rate increases but becomes weaker, the body temperature drops and coma, convulsions and rarely death ensue. The initial syndrome is what small boys might experience on smoking one of their father's cigars.

First Aid and Medical Treatment: If vomiting has not occurred spontaneously, start gastric lavage since nicotine is rapidly absorbed and if the victim has gotten too much toxin, he may quickly go into a comatose state and convulse. Give activated charcoal. If the victim is very agitated or convulsive, give diazepam. If there is evidence of excessive bronchial secretion, give atropine.

YELLOW JESSAMINE, CAROLINA JESSAMINE (all parts) rarely

While this southern belle may lull you into adoring her when you experience the large yellow, tubular flowers pouring clouds of perfume into the air, be on guard. This vine, *Gelsemium sempervirens*, is a deadly relative of the strychnine plant, from which rat poison is produced. The short-stalked, two-inch-long, lance-shaped leaves lack teeth and are produced in pairs along the wiry climbing or trailing stem. The fruit is a flattened, pointed capsule. This species should be distinguished from the jessamines of the genus *Cestrum* (p. 67).

Distribution and Habitat: This climbing, tangling, or carpet forming vine is found in the Southeast along the Coastal Plain and lower Piedmont from Virginia to Texas, and on down into Mexico and Guatemala. The ancient pre-glacial affinity of the flora of the Southeast with the flora of East Asia is shown by the occurrence in East Asia of the only other yellow jessamine.

Yellow Jessamine
Gelsemium sempervirens

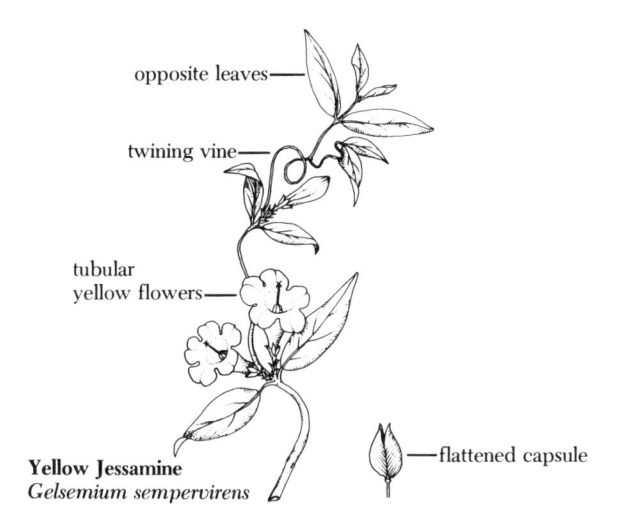

- opposite leaves
- twining vine
- tubular yellow flowers
- flattened capsule

Toxins: The toxic elements found in highest concentration in the roots and flowers are a group of bitter, strychnine-like alkaloids, the most important being gelsemine, gelseminine and gelsemoidine. The poisons appear to act by blocking the nerve endings controlling muscles, so that the vomiting response is suppressed and the ingested plant remains in the body. In folk medicine, the roots of this plant are used as a general depressant to inhibit coughing spasms in cases of pneumonia, whooping cough and asthma.

Symptoms: While all parts of the plant are poisonous, poisoning most commonly occurs when children suck nectar out of the fragrant flowers. Honey made from jessamine nectar can also contain the poison, though some of the honeybees that gathered it may die. Symptoms may begin within 15 minutes and include sweating, difficulty in speaking and swallowing, general weakness with drooping of the head and jaw, marked drooping of the eyelids, dilated pupils, muscle spasms of the arms and legs, rapid but weak pulse, depression, and rarely convulsions. In severe cases, death results from respiratory failure. Couldn't South Carolina have chosen a less deadly plant as their state flower?

First Aid and Medical Treatment: Usually the victims of these toxins don't vomit; therefore induce vomiting and administer activated charcoal. Monitor respiration. If signs and symptoms of poisoning are present give I.V. Physostigmine, 0.5 mg for children, 2 mg for adolescents and adults. Physostigmine is short acting and may be given again if symptoms reappear.

BITTERSWEET FAMILY

The bittersweet family (Celastraceae) has not been implicated in any cases of poisoning in North America, but examples of poisoning in children are known from Europe. Members of the family are recognized by their small, greenish flowers followed by a three- to five-lobed fruit, which splits open at maturity to reveal a brightly colored fleshy layer surrounding the seeds. The bittersweets *(Celastrus orbiculatus* and *C. scandens)* are climbing, woody vines with alternate ovate leaves. In the fall, the three-lobed fruit splits open its bright yellow covering to reveal the dazzling orange or red flesh-covered seed within. The persisting colorful display makes bittersweet a favorite dried flower arrangement for winter months. Also in the family is the spindle tree *(Euonymus),* common, upright, climbing or sprawling shrubs and small trees, with opposite leaves and often green twigs. Many sprawling shrub forms have leaves streaked with white. The conspicuous pinkish fruit breaks open at maturity to reveal the bright orange flesh-covered seeds inside, giving these plants two other common names: strawberry bush and hearts-a-busting. (**Plate 25.**)

Distribution and Habitat: The American bittersweet *(C. scandens)* is found in open woodlands and river edges all across eastern and central North America to Manitoba and New Mexico. The Asian bittersweet *(C. orbiculatus)* was introduced and has become

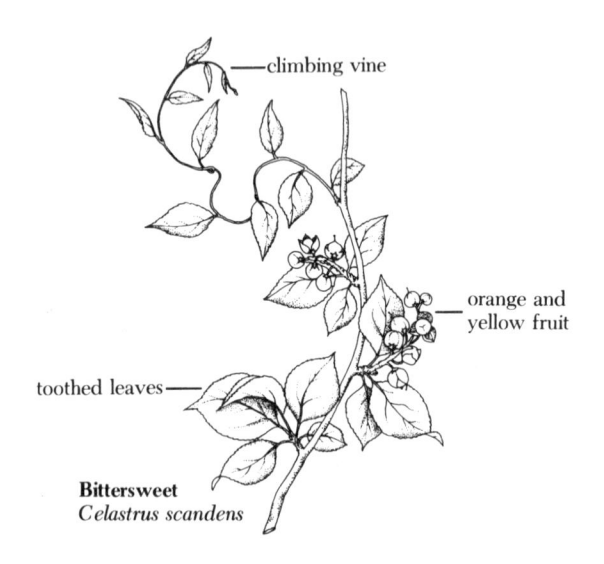

Bittersweet
Celastrus scandens

— climbing vine
— orange and yellow fruit
— toothed leaves

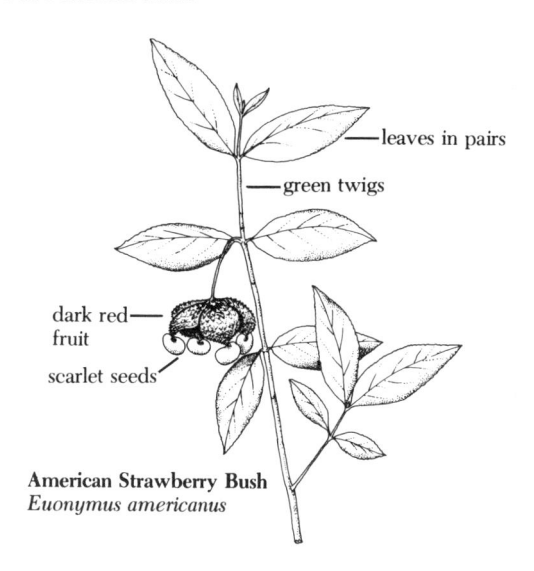

American Strawberry Bush
Euonymus americanus

- leaves in pairs
- green twigs
- dark red fruit
- scarlet seeds

widely naturalized in disturbed roadside forests, old farms and scrubby thickets. Several native species of spindle trees, the most common being *E. americanus*, occur in eastern North America, but numerous European and Asian species, such as *E. europaea*, *E. japonicus*, and *E. fortunei* have been widely planted in North America and persist in old homesites.

Toxins: All plant parts, with the possible exception of the fleshy layer around the seed, contain a bitter-tasting toxin which is not fully characterized. This toxin seems to be a glycoside.

Symptoms: Poisoning usually occurs when children eat the pretty fruits, resulting in vomiting and diarrhea. The toxin acts on the neuromuscular system, leading to weakness, coma, and convulsions.

First Aid and Medical Treatment: Treat symptoms as they arise.

COYOTILLO, TULLIDORA, CAPULINCILLO, (all parts)
GALLITA BUSH   rarely

Description: Coyotillo *(Karwinskia humboldtiana)* is a shrub or small tree packing a Mexican "mickey" more potent than a blackjack in a bordertown back alley. The distinctly stalked leaves are produced in pairs. The oval to elliptic leaves are one to three inches long, rounded at both ends and covered with short hairs

Coyotillo
Karwinskia humboldtiana

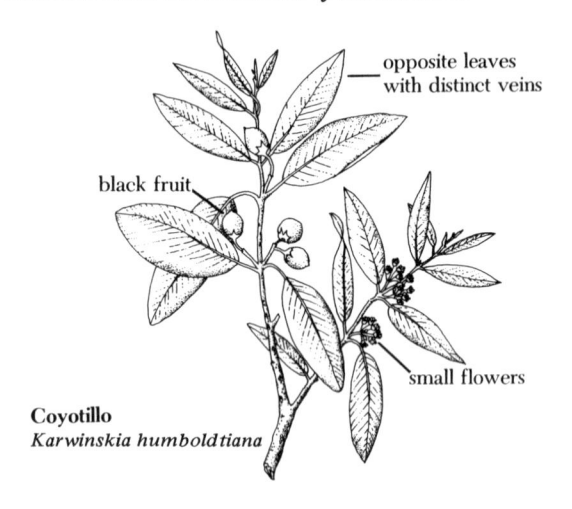

(labels: opposite leaves with distinct veins; black fruit; small flowers)

on the lower side. The small, green flowers are produced in clusters along the leafy twigs. The one-seeded fruit looks like a dark brown grape.

Distribution and Habitat: The coyotillo occurs in dry, rocky and hilly habitats from southern Texas on through Mexico and Baja California.

Toxins: The active toxin has not been identified, but other members of the buckthorn family contain glycosides.

Symptoms: While all plant parts should be considered toxic, poisoning usually occurs when children eat the fruit. Both the flesh of the fruit and the seed are toxic. The only symptom is apparently paralysis, generally of the lower limbs. The paralysis may gradually disappear or lead to death through complications in severe cases. A peculiar feature of coyotillo poisoning is the time lag between eating the fruit and the onset of the symptoms. In experiments with sheep, the signs of paralysis did not appear until more than 100 days after eating coyotillo. The foolish wild plant forager in the desert could become a walking dead man after taking coyotillo fruit as a meal. In Mexico folk medicine, coyotillo potions have been used to combat convulsions, though the paralyzing cure does not seem like much of an improvement.

First Aid and Medical Treatment: Induce vomiting and give activated charcoal. The best advice seems to be to treat the paralysis symptomatically.

117 Wild Poisonous Plants

Common Buckthorn
Rhamnus catharticus

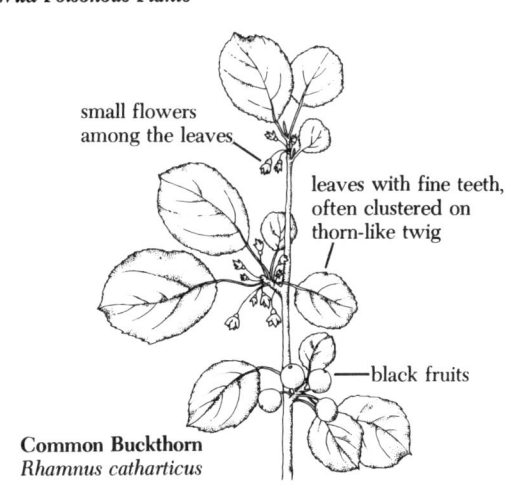

- small flowers among the leaves
- leaves with fine teeth, often clustered on thorn-like twig
- black fruits

BUCKTHORN, ALDER BUCKTHORN   (berries) rarely ☠

Description: Buckthorns in the genus *Rhamnus* are the most ordinary undistinguished shrubs and small trees and the most difficult for botany students to learn to recognize. Small, white breathing parts or lenticels are often found on the young bark. The medium-sized oval leaves sometimes have small teeth along the margin. The leaves usually alternate along the twigs, but may be produced in pairs. The twigs sometimes end in a blunt thorn. The small whitish to greenish flowers are produced in clusters among leaves. The fleshy pea-sized fruit has two to four seeds.

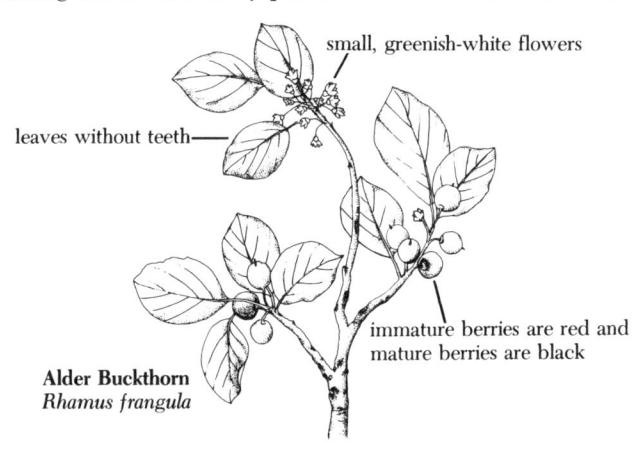

Alder Buckthorn
Rhamus frangula

- small, greenish-white flowers
- leaves without teeth
- immature berries are red and mature berries are black

The fruit may be green or red when unripe, becoming black and bitter tasting at full maturity.

Distribution and Habitat: The common buckthorn *(Rhamnus catharticus)* and the alder buckthorn *(R. frangula)* were introduced from Europe as ornamental trees and hedges, but have by now become widely naturalized. In the eastern United States, the obnoxious buckthorns are often the most common shrubs encountered in river forests and slightly disturbed moist scrubby areas. Several native species also occur in North America, such as cascara sagrada, *R. purshiana,* of the Pacific Northwest.

Toxins: The fruits contain mixtures of saponins and glycosides.

Symptoms: Eating the unripe and ripe berries may cause nausea, violent stomach pains, vomiting, bloody diarrhea and even prostration. Serious poisoning is not typical except for children. While buckthorns are only rarely lethal and the effects usually not permanent, the buckthorns are included here because they are becoming so common in our woodlands. While herbal medicine books suggest using the alder buckthorn as a laxative, its powerful action and the difficulties in distinguishing it from the common buckthorn argue against its medicinal use. In the Pacific Northwest, a laxative for relief of chronic constipation is made from the bark of cascara sagrada or "sacred bark," a name presumably given to it by early Spanish explorers. This common name may be an informative comment on the effects of their extended diet of biscuits and dried meat.

First Aid and Medical Treatment: If the patient has vomited extensively, replace fluids and electrolytes; otherwise treat symptoms as they arise.

DELPHINIUM spp., LARKSPUR

ACONITUM spp., ACONITE, MONKSHOOD, WOLFSBANE
(leaves, flowers)   rarely ☠

Since ancient times, the closely related larkspurs and monkshoods have been recognized as powerful poisons, often used to coat arrows and daggers. These plants of the buttercup family have also been used as medicine to treat neuralgia and rheumatism, but their extreme effects make them generally regarded today as too dangerous to use. Larkspurs and monkshoods are upright, one- to four-feet-tall perennial herbs with a dark, tuber-bearing,

Wild Poisonous Plants

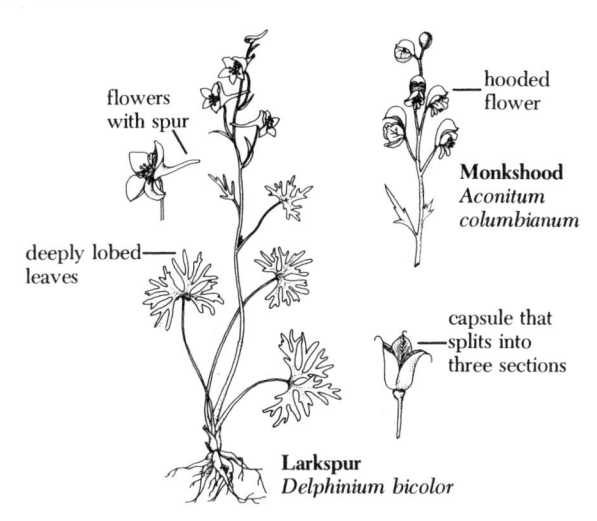

*Figure labels: flowers with spur; deeply lobed leaves; **Larkspur** *Delphinium bicolor*; hooded flower; **Monkshood** *Aconitum columbianum*; capsule that splits into three sections*

creeping rootstock, though cultivated forms may be grown as annuals. The leaves, which are produced in a cluster at ground level and then alternately along the stem, have long leaf stalks and palm-like vein patterns in the leaves. The toothed leaves are either highly lobed or divided into numerous leaflets. The showy blue, white, yellow, red or rose-colored flowers are produced in summer in a narrow spike. Unlike most other plant species, it is the sepals that are brightly colored, while the petals are small and less conspicuous. In the larkspur, the uppermost sepal has a long, nectar-filled spur, projecting backward, which serves to attract pollinating animals to the flowers. In the monkshood, the upper sepal is raised over the rest of the flower to create an appearance of a monk's hood or medieval helmet. The inconspicuous dry brown fruit splits into three to five parts at maturity to release numerous small seeds.

Distribution and Habitat: European and Asian species of monkshood, such as *A. nepellus*, and larkspur, such as *D. ajacis*, *D. grandiflorum* and *D. cheilanthum*, are commonly grown in flower gardens for their showy spikes of flowers. The blue- to purple-flowered western monkshood *(A. columbianum)* is native to mountain meadows of western North America from Canada to New Mexico, while the eastern United States has two native monkshoods which occur in rich woods and along mountain streams, the white-flowered *A. reclinatum* and the blue-flowered *A. uncinatum*. Larkspurs occur in dry rocky woods, rich woods and open habitats throughout North America, most commonly in the west.

The cultivated European larkspur, *D. ajacis*, has become naturalized as a weed on disturbed ground and cultivated fields.

Toxins: All plant parts, in particular the roots, seeds and young leaves, contain a mixture of related polycyclic diterpenoid alkaloids, such as aconitine, delphinine, and ajacine. These poisons stimulate heart nerve impulses at low doses, and inhibit nerve impulses at higher doses. Aconitine is used in medical research to disrupt the normal heartbeat so that the usefulness of new heart drugs can be tested experimentally.

Symptoms: The alkaloid is reputed to be so toxic that the sap may cause numbness or a prickling of the fingers just by picking the flowers. Poisoning may occur from eating the leaves as salad greens, the colorful flowers or other plant parts. Symptoms develop rapidly, often within half an hour, and include a numbness, tingling or burning feeling throughout the mouth, lips and throat, which may spread to the entire skin surface and to the extremities. Later symptoms include vomiting, diarrhea, confusion, general weakness, stomach cramps, muscular cramps and general body pain, and difficulty with vision. In severe cases, cardiac rhythm disturbances may lead to difficulty of breathing and convulsions, and even death within several hours.

First Aid and Medical Treatment: Induce vomiting and give activated charcoal. Administer intravenously Atropine sulfate. The initial dose for a child is 0.05 mg/kg body weight. The total adult or adolescent dose is 2-3 mg. Repeat doses of Atropine can be given at two-to five-minute intervals until the symptoms cease. Monitor heart function with electrocardiogram and treat arrhythmias (heart irregularities) as they appear.

MISTLETOE (berries, leaves) rarely ☠

The mistletoe (*Phoradendron* sp.) is most commonly encountered as a Christmas decoration where it hangs over doorways and provides an excuse for surprise kissing. Like the unwanted cad who needs this plant as an excuse for his misguided affections, the mistletoe is a parasite, taking much of its nourishment from a host tree on which it perches. Mistletoes are pale green shrubby herbs with leathery, oval, three- to five-veined leaves produced in pairs. The tiny flowers on a short spike develop into a pea-sized, fleshy white fruit with a single seed.

Distribution and Habitat: Mistletoes are more or less common parasites of oaks and other trees. Two mistletoes are common in the United States, *P. villosum* in the Pacific states and *P. sero-*

121 Wild Poisonous Plants

Mistletoe
Phoradendron serotinum

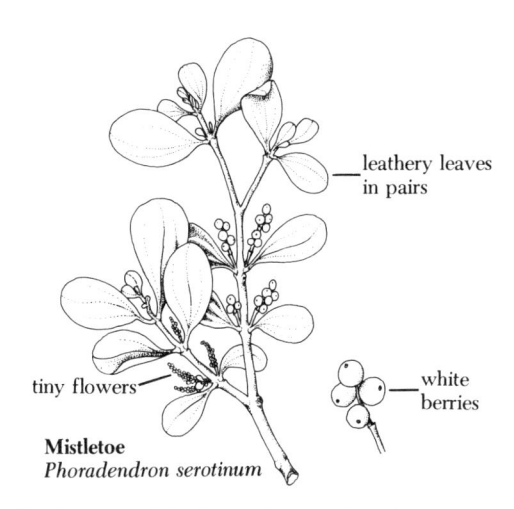

- leathery leaves in pairs
- tiny flowers
- white berries

tinum (*P. flavescens*) in the eastern and midwestern states. In tropical America, mistletoes are much more common and often have brilliant red and orange flowers.

Toxins: Mistletoes contain the toxic amines, beta-phenylethylamine and tyramine.

Symptoms: Some hours after eating the berries or drinking mistletoe tea, the victim experiences acute stomach cramps, diarrhea, and slowing of the pulse. Failure of the cardiovascular system can lead to death in severe cases.

First Aid and Medical Treatment: Monitor electrocardiogram and treat arrhythmias (heart irregularities) as they appear.

FALSE HELLEBORE, INDIAN POKE, CORN-LILY (young shoots) rarely ☠

The false hellebores are part of a small genus, *Veratrum*, consisting of ten or so poisonous species that look like wild corn plants. These tall, perennial, erect herbs (to eight feet tall) grow from a thick, short root. The major distinguishing feature of these species are their large, untoothed, oval leaves (six to 12 inches long), found in three alternating rows along the stem. The key characteristic of these leaves is their numerous lengthwise pleats. The inflorescence of the plant is a large, open, more or less branched structure at the top, bearing relatively plain-looking whitish, yellowish, or greenish flowers with six petal-like parts

False Hellebore
Veratrum viride

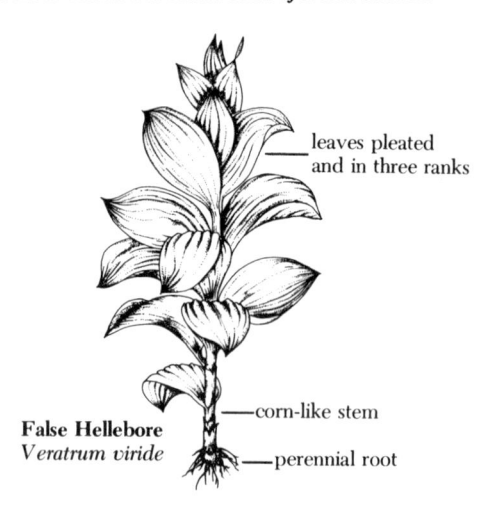

- leaves pleated and in three ranks
- corn-like stem
- perennial root

and six stamens. In the spring, the young shoots are succulent and look appetizing. However, their gastronomic temptation belies their potentially toxic impact. Aficionados of wild natural edibles sometimes mistake them for young skunk cabbage leaves, which are reputed to be edible.

Distribution and Habitat: *Veratrum californicum* is found in low, moist habitats, damp meadows, streams and swamps in mountainous areas of western North America. *V. viride* is found throughout northeastern North America into the mountains of the southern Appalachians. *V. parviflorum* occurs in the southern Appalachians from West Virginia to Georgia, and is even found in Alaska.

Toxins: The false hellebore poisons consist of a mixture of alkaloids, the most important of which are glycoalkaloids and ester alkaloids, such as veratramine. In the past, these compounds had been used both as insecticides and as medicine to treat cases of high blood pressure. In medieval Europe, the plant was used to make poison arrows.

Symptoms: After consuming any plant part, the victim experiences a painful burning or prickling in the mouth and throat, which is followed by excessive salivation, thirst, vomiting, diarrhea, dizziness, reduced urine output, muscle cramps, and shivering. The poison also acts to lower blood pressure, decrease pulse rate and lower body temperature. In severe cases, symptoms can eventually lead to generalized paralysis, convulsions and death

Wild Poisonous Plants

due to respiratory failure approximately three to 12 hours after ingestion. This plant was used by certain North American Indian tribes in a tribal ritual to choose a new chief. In this ritual, each brave would eat a piece, and whichever brave withstood the effects of the poison with the greatest composure was selected as the new chief.

First Aid and Medical Treatment: Induce vomiting, followed by activated charcoal. Be prepared to treat hypotension or respiratory failure.

LANTANA (black to green fruit) rarely ☠

The lantana shrub *(Lantana camara)* is reputed to be brewed as a tea in the West Indies as a general medicine for colds and flu. However, in North America lantana is known both as a deadly killer of cattle on rangelands and an occasional source of human poisoning. Lantana may be most easily recognized by its flat-topped clusters of small, tubular, four-lobed flowers that are first white, yellow, or pink, but change with age to darker hues of red, purple, and orange creating a multi-colored floral effect. This hairy shrub has square twigs and may have a few small spines. The leaves are produced on leaf stalks in pairs. The leaves are ovate in shape with a slightly toothed margin, and may be slightly fragrant when crushed. The pea-sized, fleshy fruit contains one seed, and is green when ripening and black when ripe.

Distribution and Habitat: This shrub is a native of the subtropical and tropical Americas in dry woodland and scrubby areas. It is

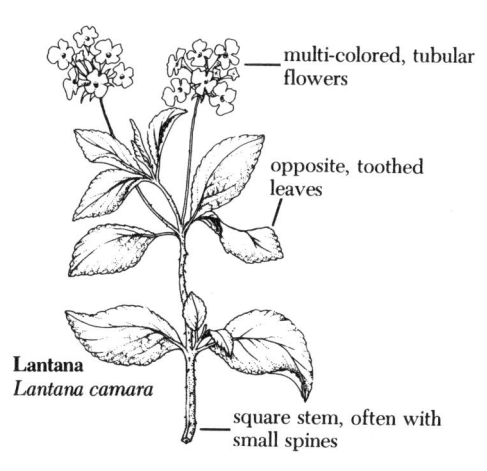

Lantana
Lantana camara

- multi-colored, tubular flowers
- opposite, toothed leaves
- square stem, often with small spines

widely planted outside as an ornamental shrub and is also grown as a potted plant. Lantana often becomes a noxious but spectacularly beautiful weed on certain rangelands in Florida, California, and Hawaii.

Toxins: The toxin is the alkaloid lantadene, and is found in highest concentrations in the immature green fruit. In grazing animals, the lantadene is altered in the liver to produce phylloerythrin, which causes photosensitization (severe susceptibility to sunburn) and jaundice.

Symptoms: After eating the fruit, the victim will experience vomiting and diarrhea, as well as weakness, slow and difficult breathing with blueness in the face, and dilated pupils with sensitivity to light, leading in severe cases to coma and death.

First Aid and Medical Treatment: Provide fluid and electrolyte replacement as needed; otherwise treat symptoms as they appear.

RELATED SPECIES (TO *LANTANA*): GOLDEN DEWDROP, PIGEONBERRY, *DURANTA*

The golden dewdrop *(Duranta repens)* looks similar to *Lantana*, but has pale blue flowers and yellow juicy fruit, with up to eight seeds and enlarged sepals partially enclosing the fruit. The golden dewdrop is a native of the American tropics, reaching Key West, but is now widely cultivated in Hawaii and the South for its flowers and attractive fruit display.

HOLLY (berries) rarely ☠

The common English holly tree *(Ilex aquifolium)* and the American holly tree *(I. opaca)* are readily recognized planted trees and Christmas decorations, with dark green spiny-margined leaves and pea-sized red berries. The cheery Christmas motif disguises the fact that the berries of these plants are dangerous if eaten. However, the trees are not as dangerous as they could be, since holly trees grow up as either male trees or female trees, so only half of the trees ever bear the poisonous fruit. The flowers are small, white and inconspicuous. The fruits have two to eight seeds and a little dark knob on the top of the fruit.

Distribution and Habitat: The American holly and other red, orange, and black-fruited native holly species occur in eastern and southern North America, often in thickets and swamps. The American holly, the English holly and other Asian species are planted in gardens as evergreen ornamentals.

Toxins: The fruits and leaves contain a mixture of the caffeine-like alkaloid theobromine, caffeine itself and glycosides. These alkaloids act to stimulate the nervous system in small doses and depress the nervous system in large doses. In South America, a local holly, yerba mate (*I. paraguayensis*), is a widely used stimulant drink. In North America, the Indians of the southeast used the yaupon tree (*I. vomitoria*) to make a hallucinogenic super-caffeine drink, which causes vomiting as a side effect. During the coastal blockade of the American Civil War, southerners rediscovered the use of mild yaupon tea as a caffeine beverage to substitute for unavailable coffee and tea.

Symptoms: Eating berries is the common way in which poisoning occurs. The victim experiences vomiting and diarrhea, with drowsiness, coma, and death occurring in severe cases. Twenty berries may constitute a lethal dose.

First Aid and Medical Treatment: If the subject is alert, induce vomiting and give activated charcoal. If the patient is drowsy, institute gastric lavage, followed by activated charcoal and cathartics. Low doses of this poison cause central nervous system stimulation which can be countered with barbiturates or Benzodiazipines. Be alert for central nervous system depression caused by higher doses of poison.

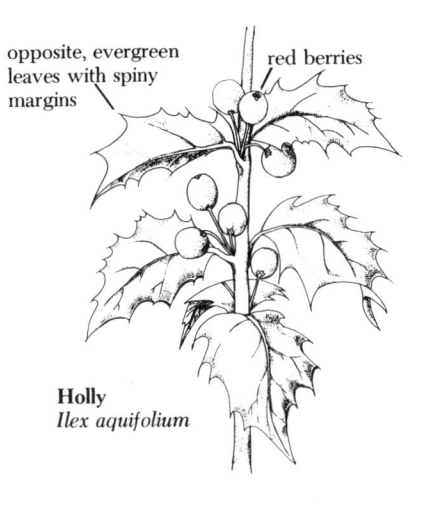

opposite, evergreen leaves with spiny margins

red berries

Holly
Ilex aquifolium

TANSY (all parts) rarely ☠

Tansy *(Tanacetum vulgare)* is a perennial European medicinal herb containing the strong-smelling oil tanacetin. This oil has been used by herbalists since the Middle Ages to kill intestinal worms and induce abortions. Any substance that can kill worms in small doses can kill you in larger doses. Cases have been recorded of deaths through severe gastrointestinal irritation and convulsions, resulting from people using excessive doses of tansy oil. Strong teas made from the plant may cause symptoms as well. Tansy is readily recognized by the unbranched, three-foot-high stems bearing leaves with deep indentations and toothed margins. The numerous flower heads in a flat-topped cluster look just like yellow buttons. All plant parts have the powerful, penetrating aroma of tansy. Tansy is commonly grown in herb gardens, and has become naturalized in disturbed, dry ground throughout North America.

WHITE SNAKEROOT (toxin via cows' milk) very dangerous ☠

In the 18th and 19th century, milk sickness was a very common disease in certain rural areas of North America. Whole villages and valleys had to be abandoned because of the prevalence of the disease. The disease was named milk sickness because it was thought the disease might be transmitted in cows' milk. Finally early in this century, scientists determined that the disease was caused by cows eating white snakeroot plants *(Eupatorium rugosum)* and concentrating the toxic compounds of that plant in their milk. Milk sickness is no longer a serious problem, now that the source of the illness has been determined. Moreover, the modern practice of pooling milk from many herds in large processing factories results in the dilution of the toxin if it is present. The only danger now comes in families which keep a single cow for home milk supply in areas in which white snakeroot is abundant.

This attractive perennial herb has an erect, usually unbranched stem up to four feet tall. The paired leaves are oval in shape, longer than broad with a tapering tip and with nine to 25 large teeth on the margin. The lower leaves may be six inches in length, decreasing in size higher on the stem. The plant produces a rounded flower bouquet at the top of the plant composed of tiny white flowers grouped into small heads. The fruit is a small, dry, flat, one-seeded achene with a whorl of feathery bristles to aid in wind dispersal of the seed.

White Snakeroot
Eupatorium rugosum

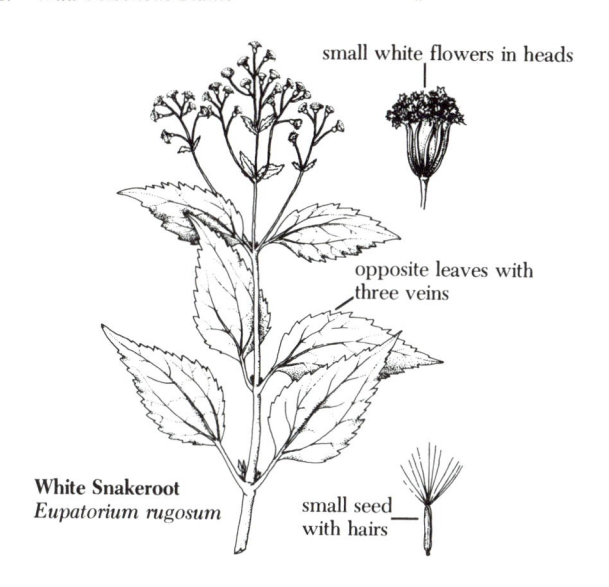

- small white flowers in heads
- opposite leaves with three veins
- small seed with hairs

Distribution and Habitat: This weed is commonly found in low moist areas, often on rich or basic soils in fields, disturbed ground, open woodlands and on streamsides, often becoming the dominant herb after logging in some areas. Snakeroot occurs throughout eastern and central North America, but is most common in the Appalachian Mountains and the north central states. Despite the widespread occurence of snakeroot, most cases of the disease have occurred in North Carolina, Indiana, Illinois, and Ohio.

Toxins: Plant parts contain a poisonous yellow aromatic oil, the most toxic component being an alcohol named tremetol.

Symptoms: Cows eating white snakeroot become sick themselves, showing sluggish behavior, stiffness in walking, trembling, gastrointestinal upset, prostration, and death. Drinking milk from a sick cow may result in weakness, loss of appetite, vomiting, stomach cramps, constipation, thirst, and muscle tremors. The victim's breath may develop the odor of acetone. In severe cases, delirium and coma may be followed by death. Mortality rates are high—10 to 25 percent—with permanent liver damage and general weakness even in recovered patients.

Prevention and Medical Treatment: The key to treatment is prevention by the identification of white snakeroot in the pasture or winter hay as the hidden source of the poison and halting the drinking of contaminated milk. Patients should be very carefully monitored for liver function and treated accordingly.

Related Species: Rayless goldenrod, jimmy weed, burrow weed. Milk sickness from tremetol poisoning can also occur from cows grazing on this perennial, bushy, two- to four-foot-tall herb *(Haplopappus heterophyllus)* of streams and irrigation ditches in the dry range country of the Southwest. The leaves are linear and distinctly sticky. Flat-topped clusters of small yellow flower heads are produced at the ends of branches.

May Apple, Mandrake (unripe fruit, leaves, roots) rarely ☠

The May apple *(Podophyllum peltatum)* is well known as an edible plant and important American Indian medicine. It is extremely easy to recognize, with its umbrella-shaped leaves, normally eight-lobed, each nearly ten inches long. This perennial herb persists by means of a thick creeping rootstock. The low stem has one leaf if the plant is not flowering. In flowering plants, a solitary nodding flower is produced on the stem between the two opposite leaves. The white, thick flower is fairly large, about two inches across, and has five to nine petal-like parts. The yellow, lime-sized fruits contain many seeds. (**Plate 26.**)

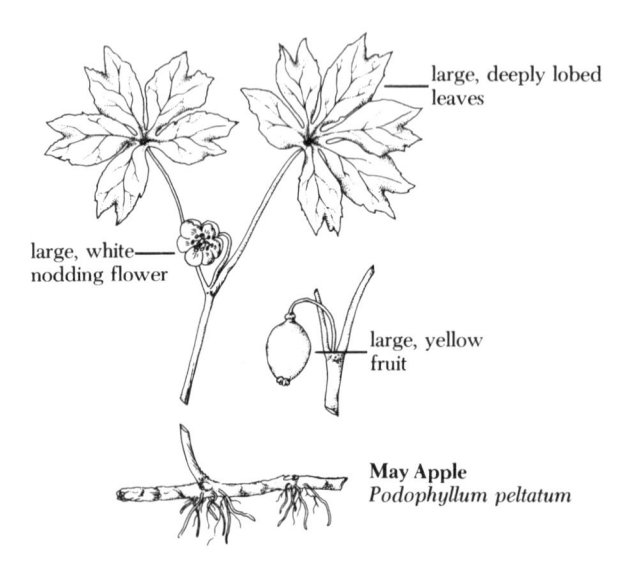

May Apple
Podophyllum peltatum

Distribution and Habitat: The May apple is found throughout North America in wet meadows, damp woods and ditches, often in large patches. It also is cultivated in gardens, where it can persist or from which it can escape.

Toxins: The toxic principle is a resin which can be extracted in alcohol and then precipitated in water. This resin, known as podophyllin, is known to be a mixture of lignans and flavonols, the most important being the glycoside podophylotoxin. The resin used to be prepared in quantity for use as a laxative and liver tonic. It has also been used to a limited degree to treat venereal warts, prompting some research on its potential as an anticancer drug. The poison apparently acts to interfere with cell division.

Symptoms: While the ripe fruit is apparently edible, the unripe fruit, leaves, and roots are all toxic. Most cases of poisoning occur through the preparation of May apple resin as a folk medicine and during the treatment of venereal warts. Inflammation of the eyes and nose can result from contact with the sap. Following eating the plant or about 12 hours after prolonged contact during a folk-medical treatment for warts, the victim may experience vomiting, severe stomach pain and diarrhea, and even weakening of the pulse, poor circulation to the extremities, a depressed central nervous system, low blood pressure, a low blood count, general dizziness, coma, and death. Even after the major symptoms have subsided, the victim may show decreased reflexes.

First Aid and Medical Treatment: If patient is alert and hasn't vomited, induce vomiting and give activated charcoal. Treat symptoms of diarrhea if they become severe. If podophyllin was used as a topical folk medicine, wash it off thoroughly. In severe cases of May apple poisoning it has been reported that charcoal hemoperfusion was beneficial.

Related Species: The blue cohosh *(Caulophyllum thalictroides)* is a perennial herb of rich woods and mountains in eastern North America. This perennial herb has a single large triangular leaf, divided into many leaflets, and a cluster of pea-sized, dark blue seeds on a central stalk. The plant contains alkaloids and glycosides, bitter in taste, which can cause stomach cramps when eaten.

MOONSEED (all parts) rarely ☠

Looking like a grape vine in just about every way, the moonseed *(Menispermum canadense)* kills by fooling the unwary. This climbing vine has leaves with long stalks and three to five large, rounded lobes, in contrast with grape leaves, which usually have many

130   Poisonous Plants and Mushrooms of North America

Moonseed
Menispermum canadense

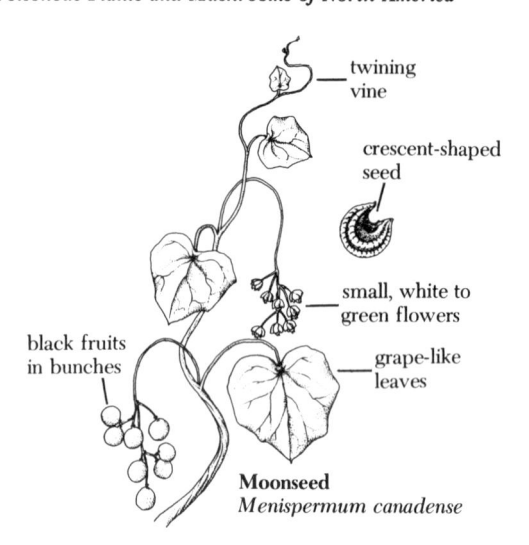

(Labels: twining vine; crescent-shaped seed; small, white to green flowers; black fruits in bunches; grape-like leaves)

small teeth along the margin. The fruits are black in color and grape-like in size and shape. However, the fruit has a single, large grooved, half-moon-shaped seed, rather than the numerous smaller seeds of a grape.

Distribution and Habitat: The moonseed is an infrequent inhabitant of thickets and moist woodlands in eastern and central North America.

Toxins: Toxins in moonseed have not been identified. It seems likely that moonseed plants possess the isoquinoline alkaloids found in other members of the moonseed family.

Symptoms: Cases of fatalities in children have been reported. The severe symptoms are likely to be those associated with alkaloid poisoning: convulsions, muscle tremors, coma, and cardiac and respiratory failure.

First Aid and Medical Treatment: Induce vomiting and give activated charcoal. Treat symptoms as they occur.

PART V
Poisonous Mushrooms

Mushroom hunting in Europe, Asia and Central America has a long history. Mushrooms have been eaten in diverse societies for two reasons. The first is that some of these fleshy fungi are flavorful gourmet treats that come in a variety of tastes and textures. We now know they have real nutritional value, containing small amounts of protein (3% by weight), some carbohydrates, lipids, minerals and vitamins. They also are low in calories, about 120 calories per pound, making them desirable in modern diets. A second reason that mushrooms are eaten has to do with the fact that several species contain substances which produce hallucinations, and hence have been used for centuries to produce "highs" or religious experiences. There is even a specialty area, called ethnobotany, which studies the use of hallucinogenic fungi and plants by diverse ancient societies.

Wild mushroom hunting in Europe is a major hobby bordering on a passion, and it is also a profitable business. Truffles, a valuable delicacy, are hunted by trained dogs and trained pigs that use their superb sense of smell to snuffle truffles. A good truffle hound might bring as much as $2,000 in the open market. In the United States, mushroom hunting until the 1960s was largely limited to European immigrants, but recently, with the increased interest in natural foods and a new sophistication in American tastes, there has been a great upsurge in mushroom hunting, which not only provides delicacies for the table but also is an excellent recreational activity and hobby. However, life involves a series of risk-benefit and cost-benefit decisions, and the pursuit of wild mushrooms does have risk and potentially high costs in terms of health and survival because many of the 300 or so species of mushrooms collected can be toxic to some degree and ten to 15 species can be deadly.

INCIDENCE OF MUSHROOM POISONING

In Europe where wild mushroom hunting is common, there is a high incidence of mushroom poisoning, and in Germany in one year there were 200 recorded deaths caused by eating toxic fungi. The data regarding mushroom poisoning in the United States and Canada are at best sketchy, and only a small portion of actual cases are reported to Poison Control Centers, since such reports are not mandated by law. The National Clearinghouse for Poison Control Centers has only limited information, since those cases treated by private physicians and hospital emergency rooms are often not represented in the data. Thus, according to the poison control clearinghouse, there were 1,508 verified cases of mushroom poisoning reported in 1975. Of these only 17 victims required hospitalization and there were no reported lethalities. The majority of the victims in this report were children under five. Other epidemiological studies estimate that there are two or three deaths a year that can be attributed to eating poisonous mushrooms. The U.S. Department of Agriculture Poison Fungus Center, using data from only eight states, had reports of 105 poisonings and two deaths. However, a combined effort in 1972–75 by the Rocky Mountain Mushroom Club, the Colorado Poison Control Center and state medical facilities found that there were 50 confirmed cases of mushroom poisoning per year in that state alone. According to Dr. Lot B. Page, Chief of Medicine at Newton-Wellesley Hospital in Massachusetts and a specialist in medical mycology, ". . . it is likely that the true frequency for the nation as a whole is much higher than the reports suggest." The number is also on the rise as more people escape to the great outdoors.

One of the major problems in detecting mushroom poisoning is that most physicians are inexperienced and minimally trained in this area because it represents only a minor health problem in terms of number of victims and deaths. Another reason for underestimating the incidence of fungi poisoning is that in most cases of toxic mushroom ingestion, the main symptoms mimic the gastrointestinal upsets associated with bacterial or viral gastroenteritis (vomiting, cramps, and diarrhea), and these symptoms are usually of brief duration (several hours) and recovery is spontaneous. Another problem with mushroom poisoning reporting is that the more deadly species contain toxins that don't produce their effects for several hours or even a few days after ingestion;

hence the establishment of cause and effect is obscured. One group of mushrooms that can cause serious kidney damage has a delay of up to three weeks.

MUSHROOM TOXINS

In recent years, there has been some progress in mushroom toxicology. Many of the toxins, particularly the deadly and hallucinogenic substances, have been isolated, characterized and studied in some detail. However, there are still quite a few as yet undetermined toxins. Furthermore, the determination of those species that are or may be poisonous is often very uncertain. In preparing this book, we went through every available field guide, and examined much of the recent scientific literature, which revealed wide discrepancies and differences of opinion. Part of the reason for this confusion is that names of both genus and species are being changed constantly to fit varying systems of classification. Secondly, many of the manuals are chatty personal accounts, e.g., "I have heard of people becoming ill after eating this mushroom, but I personally found it quite tasty and without ill effect." How could a respected expert come up with such a conclusion? Part of the cause of such variability is that the toxic content may vary with season and locale. Also, subspecies or races of a mushroom may vary so much that even an expert can become confused. Finally, there are marked individual sensitivities to mushroom toxins. The biggest problem for the new mushroom hunter is how to identify any particular mushroom with certainty. Unfortunately, exact species verification for many mushrooms often requires microscopic examination and chemical tests, which reduces the value of field characteristics as criteria for identification. Thus, if you are learning about mushrooms and want to go collecting, go with experts. Local mycological clubs in most cities and universities sponsor frequent trips. This field guide will provide descriptions and drawings which point out key field identification features of the more common toxic and deadly mushrooms. Any mushroom that looks like one of those described herein should be avoided. Just remember, "There are old wild mushroom eaters and bold wild mushroom eaters, but very few, if any, old and bold wild mushroom eaters." According to Dr. A.H. Smith, a well known mycological expert, "To anyone who isn't an expert, I suggest that he limit his mushroom hunting to the shelves of the supermarket."

MYTHS ABOUT DETERMINING EDIBILITY

Contrary to some popular beliefs, not all poisonous mushrooms blacken or tarnish silver or cause a clove of garlic cooked with them to blacken. Another erroneous belief is that a mushroom is safe to eat if it can be peeled. Not so! Some collectors have the mistaken idea that all mushrooms growing on wood are edible, but several poisonous species, including the deadly small brown *Galerina autumnalis*, grow on wood. It has also been suggested that mushrooms that show signs of being munched by animals are safe, but don't believe it. While it is true that some mushroom toxins can be removed by parboiling in several washes of water, there are still many cases of poisoning from eating parboiled mushrooms. Likewise the myths that field mushrooms are edible and that mushrooms that grow in clusters are safe can lead unwary gourmets to the emergency room.

PRECAUTIONS TO BE STRICTLY OBSERVED

The following list is provided by the Forest Service of the U.S. Department of Agriculture:

1. Be very sure of your identification. Keep different kinds of mushrooms separated while collecting, make spore prints, compare the collections carefully with descriptions in good field guides and ask an expert for confirmation. If you feel that you must collect and eat wild mushrooms, pick out a few selected species whose characteristics are very well known to you and leave the rest alone.
2. Save a few fresh mushrooms from the pot. If you make a mistake, an expert can identify the mushrooms and you will have a much better chance of getting the proper treatment.
3. Do not eat mushrooms raw. Some are all right raw, but others have toxins that are either destroyed by cooking or are volatile and boil off.
4. Eat only mushrooms in good condition. Spoiled mushrooms that would have been good when fresh can cause illness.
5. Eat only one kind at a time, eat a very small amount the first time and never eat large amounts. In small quantities, mistaken identification or an unsuspected allergy will have less disturbing consequences. If several poisonous kinds are eaten to-

gether, the symptoms will be confusing and treatment may not be as effective.
6. Do not feed wild mushrooms to anyone but healthy adults. Children, the very old, and the chronically ill can be poisoned by mushrooms that are harmless to healthy adults.
7. Do not experiment. Don't be the one to demonstrate that a previously untested mushroom from a given region is edible, only to discover that the mushroom is deadly.
8. If you suspect poisoning, immediately call the local poison control center. Save any mushrooms, cooked mushroom dish, or vomit so an expert can identify the mushroom.
9. Avoid little brown mushrooms (LBMs) and any pure white mushrooms since white mushrooms could be the dangerous *Amanita*.

CLASSIFYING MUSHROOMS BY TOXIN

Several different authors have tried to group poisonous mushrooms by their toxins but this is difficult since a number of poisonous mushrooms don't fit the usual categories. To overcome the problem of classification, we have added a category of ill-defined poisons and further add a warning that most poisonous mushrooms cause gastrointestinal upset but may also have other effects. Accordingly, we have grouped the mushrooms as follows: In the first group are mushrooms that are deadly (they include several species of *Amanita*, the false morels and some little brown mushrooms of the genus *Galerina*). Be warned that some mushrooms not in this category have been responsible for deaths, but 90 percent of the deaths associated with mushroom ingestion can be attributed to this first grouping. The second group are mushrooms that contain the poison muscimol, mainly *Amanita muscaria* and *A. pantheria*. The third group are mushrooms that contain the poison muscarine and produce the so-called "toggle-switch" intoxication, mainly mushrooms of the genera *Clitocybe* and *Inocybe*. The fourth group are the so-called magic mushrooms that contain hallucinogenic compounds and belong to the genera *Psilocybe*, *Conocybe*, *Panaeolus*, and *Gymnopilus*. The fifth group includes mushrooms that are only toxic when alcohol has been consumed and they are in the genus *Coprinus*. In the sixth grouping are mushrooms that cause gastrointestinal upsets ranging from mild to severe. There are so many mushrooms that fall into this category that we have only included a few of the more common ones. The last category is a mixed bag and we have only included a few mushrooms whose toxins are somewhat ill-defined.

I. DEADLY OR POTENTIALLY DEADLY MUSHROOMS

DEADLY *AMANITAS* (all parts) deadly delayed reaction ☠

The name *Amanita* was provided by the Greek philosopher-physician Galen after mushrooms found on Mount Amanus. Within the genus are well known poisonous species and a few known edible species. However, our knowledge about certain possibly poisonous *Amanita* species is incomplete. Different sources are not in agreement so to be on the safe side we are including all *Amanita* suspected or reported to contain the deadly poison amatoxin.

The destroying angel, *Amanita virosa*, is a pure chalky white killer, widely distributed in eastern North America and only rarely in the far West. It has a wide conical cap when young, but this cap becomes less convex and ultimately flat as the fungus ages. The cap is sticky when wet and its margins are smooth. The half-inch-thick white stalk (three to eight inches tall) is tapered toward the cap, and there is a well developed but tattered skirt just below the cap. A sac-like cup or volva is found at the base of the stalk and may be covered by fallen leaves or soil litter, and thus escape observation. (**Plates 27 and 28.**)

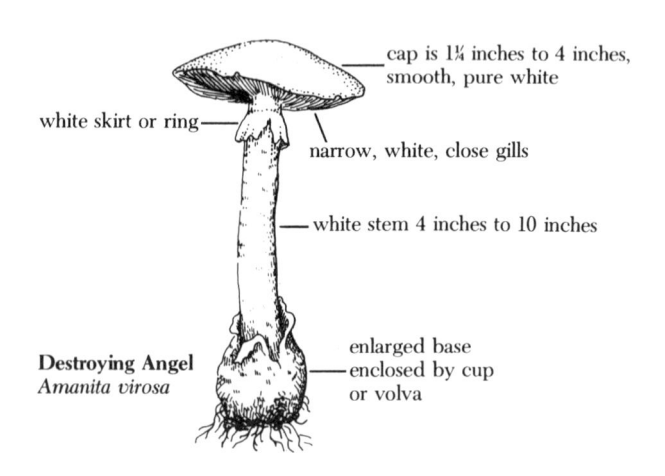

Destroying Angel
Amanita virosa

- cap is 1¼ inches to 4 inches, smooth, pure white
- white skirt or ring
- narrow, white, close gills
- white stem 4 inches to 10 inches
- enlarged base enclosed by cup or volva

Distribution and Habitat: Widely distributed in mixed woods or hardwoods throughout North America. Though some can be found in the late spring and summer, they are most common in the fall (to November), usually growing as solitary individuals or small scattered groups.

There are three other deadly large white *Amanitas* (also called destroying angel) that look so much like *A. virosa* that they can hardly be distinguished by their field characteristics. These are *A. verna*, *A. bisporigera* and *A. ocreata*.

Distribution and Habitat: *A. verna* is very common during April, May and June in wet forests of the Pacific Northwest, particularly under birch or aspen. They are also found in mixed woods in the Great Plains. *A. bisporigera*, a smaller, slender-stemmed species, is more common in the Southeast, in oak forests from midsummer to mid fall, but also occurs elsewhere. It is rarely found on the Pacific west coast. *A. ocreata* is found in the Southwest under hardwoods during the winter and early spring.

DEATH'S ANGEL, DEATH'S CAP (all parts) most deadly ☠

The most potent killer mushroom in North America is the deadly *Amanita phalloides* which is a common European species now introduced in a few places in the New World. This mushroom contains two and a half times as much toxin as the other deadly *Amanitas*. A single mushroom can be lethal to an adult; on one occasion, a child died after consuming only one-third of a death's cap. *A. phalloides* has a satin-smooth, broad, convex cap (two to six inches across) that ranges in color from pale yellow-green to olive to dark brown. Initially the cap is whitish, moist, and conical, but as the mushroom ages it flattens and may even turn up at the edges. The cap has inconspicuous radiating hairs, but the margin is smooth. Near the cap, the smooth white stem is encircled by a conspicuous hanging skirt. The stem (4 to 8 inches tall) widens toward the base, where it is engulfed by a sac-like cup or volva. Some people report that *A. phalloides* has a disagreeable odor.

Distribution and Habitat: Uncommon, under trees, particularly European-introduced species, in the summer and fall.

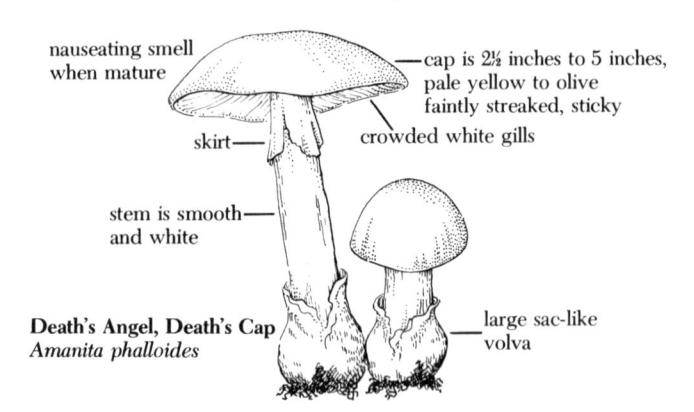

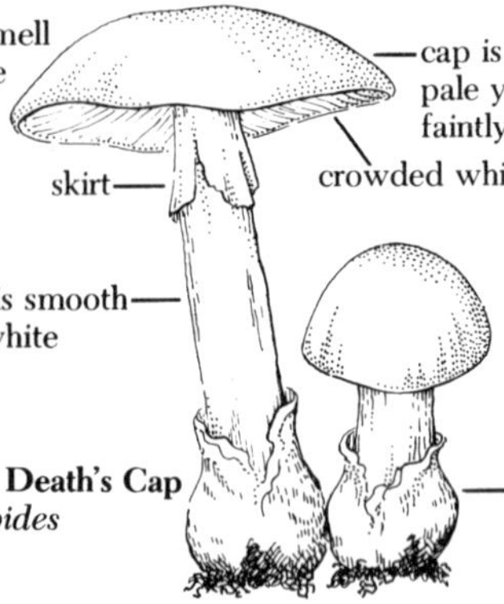
nauseating smell
when mature
—cap is 2½ inches to 5 inches,
pale yellow to olive
faintly streaked, sticky
crowded white gills
skirt—
stem is smooth—
and white
large sac-like
volva

Death's Angel, Death's Cap
Amanita phalloides

Toxins: Several groups of toxins, particularly the amatoxins (which are octapeptides), are responsible for human poisoning. These toxins selectively damage liver, intestinal, and kidney cells by irreversibly inhibiting RNA polymerase II, thus blocking the cell's ability to make new proteins. This not only causes cell death but also prevents reproduction of new cells to repair the damage. High concentrations of the toxins reaching the kidneys cause destruction of renal tubule cells. The effects of amatoxins are further enhanced because some of the poison enters the bile and is dumped right back into the intestine (via the bile duct) where it is again absorbed, producing further cellular mayhem. These toxins are not broken down by cooking and are reported to be present even in dried mushrooms stored for many years.

Symptoms: Symptoms caused by amatoxins appear from six to 24 hours after consumption. The onset is sudden: severe abdominal cramping pain, nausea, vomiting, and diarrhea may last 24 hours. In very severe cases, there is little or no urine production due to kidney tubule damage caused by the action of amatoxin. The spasms, pain, and vomiting may wax and wane, and even apparently go away for a day or so. After the acute intestinal phase and false recovery the patient gets sick again. Within one to five days jaundice (yellowish color to skin and eye whites) may give warnings of severe liver cell destruction. Depending on the severity of toxicity there may be protein in the urine, decrease in blood sugar and damage to blood vessels, leading to tiny hemorrhages in the skin. Poisoning victims may proceed into a coma, terminal convulsions with death occurring four to seven days after ingestion. Mortality is reported to be from 30 to 50 percent. Even in survivors, recovery is slow—about four weeks—and is accompanied by liver enlargement and considerable pain. The Roman Emperor Claudius, stepfather of Nero, was supposedly poisoned by *A. phalloides* to open Nero's way to the throne.

First Aid and Medical Treatment: There is no first aid available. By the time the patient presents symptoms the toxins are absorbed and clearing the gastrointestinal tract is futile. There is also no currently accepted treatment for amatoxin poisoning. Call your poison center for advice since some treatments are being clinically tested. Provide supportive care until you can get expert advice from the poison center mushroom intoxication consultant.

DEADLY LITTLE BROWN MUSHROOMS (all parts) deadly

Three members of the genus *Galerina* are small, brown mushrooms that contain deadly amatoxins, one of the two groups of deadly cyclopeptide poisons that wreak cellular mayhem in the liver and kidneys by disrupting protein synthesis. The toxins of these mushrooms are particularly insidious since symptoms may not occur for a day or so after eating them; consequently, there is no chance to vomit up these toxins before they are all absorbed. Furthermore, it is impossible to clearly classify these mushrooms from just their field characteristics. Fortunately, they are small and don't attract much attention from mycophagists (mushroom

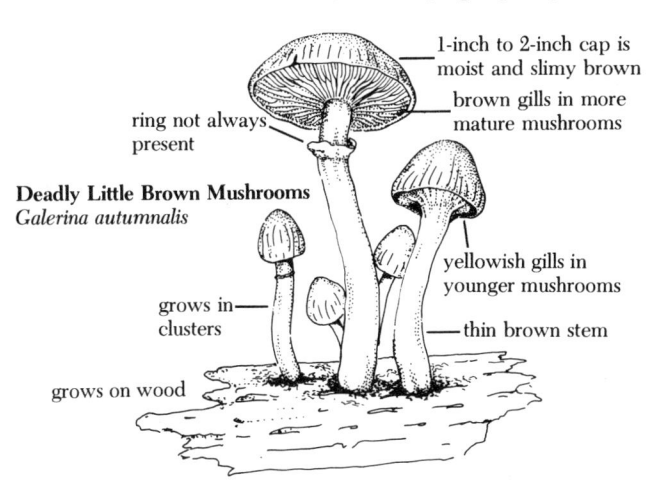

Deadly Little Brown Mushrooms
Galerina autumnalis

- 1-inch to 2-inch cap is moist and slimy brown
- brown gills in more mature mushrooms
- ring not always present
- yellowish gills in younger mushrooms
- grows in clusters
- thin brown stem
- grows on wood

eaters). However, they may attract children and it would be prudent to teach them to avoid eating any very small brown mushrooms that may be *G. autumnalis*, *G. marginata*, or *G. venenata* because death rates for these species may be as high as 60 percent. Despite the old wives' tale that mushrooms growing on wood are non-toxic, *G. autumnalis* grows very nicely on logs and stumps.

Description: *G. autumnalis* has a one- to two-inch-wide, knobbed, sticky or slimy yellow-brown to dark brown cap. The gills, which are yellowish, turn brown on maturity. The thin brown stem is often, but not always, encircled by an inconspicuous white skirt. This species is often found in clusters on decaying wood. These mushrooms are notoriously variable in size and form. (**Plate 31.**)

G. marginata has a one- to two-inch-broad, knobbed and viscous cinnamon-brown cap, while *G. venenata* has a half-inch to inch-and-a-half slimy, smooth, cinnamon-brown cap that fades as it ages. The one- to two-inch tall stem is brown and often, but not always, has a skirt up near the cap. *G. vennata* is relatively rare.

Distribution and Habitat: All three are gregarious, tending to form clusters from early summer to fall. They are particularly abundant if there has been a lot of rain. *G. autumnalis* grows on or near decaying wood (stumps and fallen logs), while the other two are found on lawns, in meadows or in open woods.

Symptoms: There are at least five amatoxins that produce delayed reactions by destroying the cell's nucleolus, the prime source of RNA, needed for the synthesis of protein. Since the toxin is absorbed by the intestines and transported directly to the liver via the hepatic portal system, damage in the liver can be extensive. Amatoxin also effects kidney cells and may also damage and kill cells of the intestinal lining. The earliest symptoms may appear six hours after ingestion, but may be delayed for as long as 30 hours or more. The longer the delay, the worse the chances for survival. Initially, the victim experiences painful severe abdominal cramps, nausea and vomiting. This is followed by bloody diarrhea. The damaged liver is enlarged and tender, and with impaired liver function there is jaundice and hypoglycemia (low blood sugar). There may be a collection of fluid in the lungs, and the hypoglycemia can be severe enough to cause confusion, coma and death. The intensity of the symptoms may wax and wane, with the peak disruptions occurring three to four days after onset. Deaths from severe *Galerina* poisoning may be as high as 60 percent.

First Aid and Medical Treatment: See *A. phalloides*, p. 139.

Deadly Conocybe (all parts) rarely ☠

Another of the very dangerous little brown mushrooms is *Conocybe filaris*. This small, seemingly innocuous mushroom packs a deadly punch, since it has been reported to contain the same poisons as the destroying angels of the genus *Amanita*. It has a brown cap that ranges from a quarter of an inch to an inch across. The striate cap may range from conical to convex in young specimens to flat with a raised central knob in older specimens. Its off-white to rusty-colored close gills are notched. The stem is yellow-orange to brown, stands only a half to one and a half inches tall and is encircled by a movable ring. BEWARE OF LITTLE BROWN MUSHROOMS.

Distribution and Habitat: This little brown assassin can be found from August through October across much of North America in scattered clusters on lawns, on moss or even wood litter. In some guides this mushroom is called *Pholiota filaris*.

Symptoms and Medical Treatment: See p. 138.

False Morels (all parts) possibly ☠

The true morels are among the best known mushroom delicacies sought after by collectors. Unfortunately, there are several look-alike cousins, known as false morels, that can be quite deadly. There are two species of *Gyromitra*, *G. esculenta* (also called *Helvella esculenta*), the most common, and *G. infula* (also called *Helvella infula*), that have caused a number of confirmed deaths. These mushrooms contain a somewhat water soluble, volatile hydrazine-type poison called gyromitrin which can cause degener-

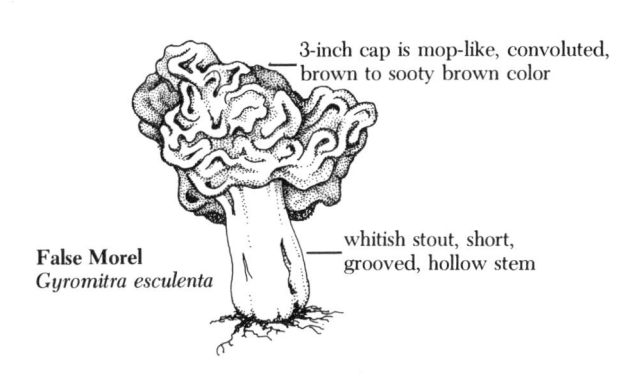

False Morel
Gyromitra esculenta

3-inch cap is mop-like, convoluted, brown to sooty brown color

whitish stout, short, grooved, hollow stem

False Morel
Gyromitra infula

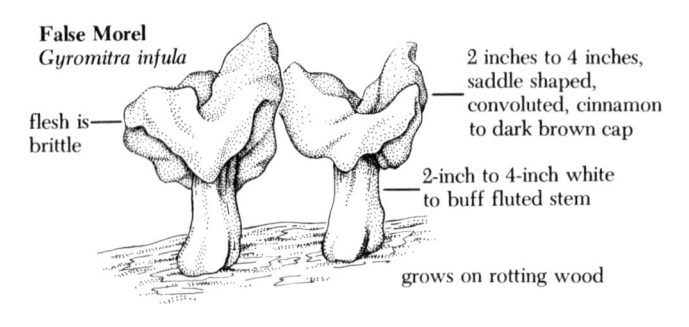

- 2 inches to 4 inches, saddle shaped, convoluted, cinnamon to dark brown cap
- flesh is brittle
- 2-inch to 4-inch white to buff fluted stem
- grows on rotting wood

ation of the liver. One possible reason for the confusion regarding its toxicity is that some of the gyromitrin may dissolve or vaporize during cooking in several changes of water. However, there are recorded fatalities even after eating cooked *G. esculenta*, so it is best that one learns to recognize and avoid this potential killer.

Description: *G. esculenta* stands up to six inches tall and has a convoluted cap three inches in diameter that looks like brain tissue. The ravelled, knot-like cap is dark brown to sooty brown in color and is firmly attached to a hollow, thick (half as thick as the cap), short white to light brown stem that is often grooved. Some authors describe this mushroom as mop-like in appearance. *G. infula* is less common, has a convoluted cap that ranges from cinnamon to dark brown and is folded back in such a way that it is saddle-shaped. Its stem is smooth and is light purple in color. *G. infula* is more common in the West than the East. Both are to be avoided. The fabled false morel, *G. brunnea*, is also reported to be poisonous. (**Plate 30.**)

These fungi are found under conifers in sandy soils or on rotting wood from early to late spring (April, May and early June). Be particularly careful of *Gyromitra* whose caps are showing signs of decay, since some reports indicate those are the most toxic. Human lethal doses are ten to 50 mg of gyromitrin/kg body weight, but children seem particularly sensitive to this toxin.

Symptoms: The reaction is of the delayed type, and symptoms appear two to eight hours after ingestion, more usually six to eight hours. Symptoms include severe cramping, upper abdominal pains, nausea, severe vomiting, diarrhea, fatigue, and jaundice a day or two after poisoning. In severe cases, the victim shows pulse irregularities, shortness of breath and a temperature, and may become delirious, convulse, and die. Death is mainly due to damage to the liver. Lethality figures vary, but some report death rates of up to 15 percent. Acute symptoms usually last two to three days and death, if it occurs, will be within two weeks after ingestion. Deaths in the United States and Canada are rare.

First Aid and Medical Treatment: Get the patient to the hospital quickly. Generally, treat symptoms as they occur. Have supportive therapy available. Vitamin B_6, pyridoxine, has been used with some success since it has been used to treat poisoning with other hydrazines. The recommended dose of vitamin B_6 (pyridoxine HCl) is to infuse 25 mg/kg body weight over 3 hours.

NEOGYROMITRA GIGAS   (all parts) possibly deadly ☠

Another poisonous morel look-alike is the large-stemmed, king-sized mushroom *Neogyromita gigas*. Although parboiling will probably remove most of the toxin, caution is urged.

Description: *N. gigas* is quite distinctive in form. It is large, having a yellow to tan, highly convoluted, brain-like cap that is two to ten inches across and hangs down over a very thick, white, folded and convoluted stem.

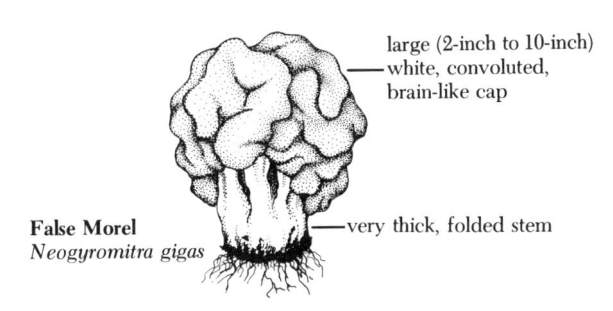

large (2-inch to 10-inch) white, convoluted, brain-like cap

False Morel
Neogyromitra gigas

very thick, folded stem

Distribution and Habitat: *N. gigas* is rather common in the spring and early summer in coniferous woods in the Pacific Northwest and the Rockies.

Symptoms: Same as *Gyromitra* (above).

HELVELLA LACUNOSA   (all parts) possibly deadly ☠

Helvella lacunosa is another false morel that contains the potent liver poison gyromitrin which is described in detail on page 141. "Helvella," which translates to "small pot herb," is a misnomer,

False Morel
Helvella lacunosa

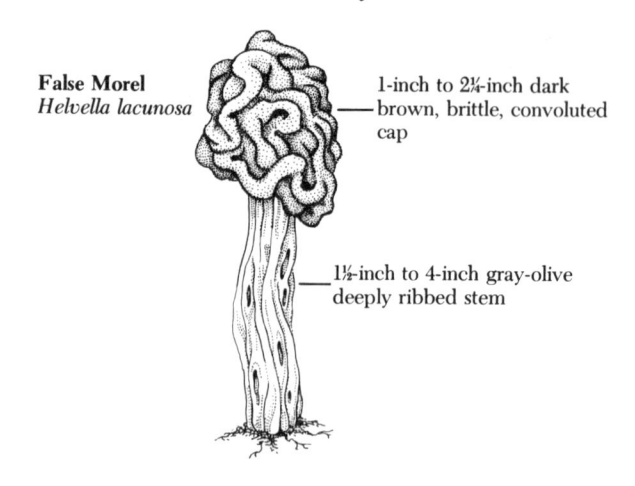

1-inch to 2¼-inch dark brown, brittle, convoluted cap

1½-inch to 4-inch gray-olive deeply ribbed stem

since it can be deadly, although cooking in several changes of water might dissolve out some of the toxin. This mushroom, particularly older specimens, should be avoided.

Description: *Helvella lacunosa* stands about four inches high. Its one- to two-inch-broad cap is ridged, pitted and lobed (two to four lobes). The cap color ranges from dirty white to sooty gray-black, and it is roughly saddle-shaped. The hollow cylindrical gray-white stem has many longitudinal depressions and ribs and attaches to the cap only at its apex.

Distribution and Habitat: *H. lacunosa* is found in the fall rainy season under conifers, e.g., Douglas fir, or under alders. It thrives particularly well in burnt-over areas.

II. MUSHROOMS THAT CAUSE PARASYMPATHETIC STIMULATION

FLY AGARIC   very rarely ☠

The fly amanita or fly agaric *(Amanita muscaria)* is a large, attractive, colorful toadstool that has been used by artists to decorate children's picture books, toys and even table bric-a-brac. Its showy and attractive features have been used by the Disney Studios as a shelter or seat for gnomes, fairies, elves and wee people. Its color is variable, orange, yellow, white, or rarely red, with the center orange-red and covered with white or cream-colored wart-

145   Poisonous Mushrooms

Figure labels:
- western species orange to blood red cap
- eastern species yellow cap
- margins striate
- skirt
- concentric rings
- bulbous base
- white to creamy warts
- 7 inches to 10 inches
- gills crowded, white
- 3-inch to 6½-inch stem, white, covered with silky hairs

Fly Agaric
Amanita muscaria

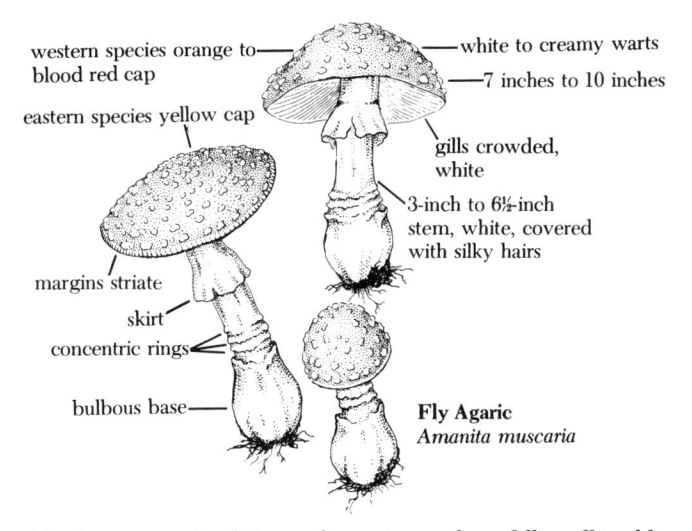

like fragments of veil tissue; the warts may have fallen off in older mushrooms. When the mushroom is young, it is spherical or ovate, and as it matures it forms a convex umbrella with slightly striated edges which flattens and then forms a flat or even slightly concave dish, whose white or creamy gills are apparent on the underside. The four- to eight-inch-tall stalk tapers slightly upward, and about two-thirds of the way to the cap there is a skirt-like ring or annulus. The remnants of the volva or cup leave a number of collar-like scales ringing the lower part of the stem. Unlike the death's angel, there is not a separate volva. (see cover, mid left and **Plate 29.**)

A. muscaria has a longstanding widespread history of use in superstitious, religious and mystical practices among primitive peoples because of its ability to induce visions and narcosis. There has been extensive research into the use of *A. muscaria* over time by ethnobotanists, and the English word berserk has its origins from a clan of Vikings who used this fungus for ceremonial intoxication. The Mayans used it for religious ceremonies, and some controversial authors suggest it played a role in early Christian beliefs. References are made to fly agaric in the story of Jason and the Golden Fleece and in *Alice in Wonderland*.

Distribution and Habitat: Fly agaric can be found throughout North America in open deciduous and evergreen woodlands, where it is often seen under birches, pine trees, aspens or even bramble bushes. It can also grow on partially cleared land and even roadsides, preferring poor soil. Fly agaric typically grows in patches or fairy rings, though sometimes it grows singly. In most areas it can be found from late spring until the fall, but along the Pacific coast it can be found during the winter months. This mushroom is often surrounded by a circle of dead flies who have sipped the virulent juices of the cap; indeed, this mushroom's juices were originally used in milk as a fly killer.

Toxins: There are a number of toxic substances in all parts of this mushroom. One, ibotenic acid, in the body is converted to its most active ingredient, muscimol, a potent stimulant of smooth muscles, glands and other secretory tissue. Also present are variable concentrations of mycoatropine, an atropine-like alkaloid which can produce excitement, fearlessness, hallucinations and even narcosis and convulsions. In Europe, estimates have been made that ten mushrooms constitute a fatal dose, though in North America this mushroom is rarely lethal.

Symptoms: Muscimol poisoning produces a spectrum of alternating symptoms which some authors call toggle-switch intoxication. As quickly as 15 to 30 minutes after ingestion, the victim may exhibit symptoms of confusion, sleepiness, visual disturbances, muscle spasms, and sometimes even convulsions. Later symptoms include dilated pupils and a dry mouth. Usually the symptoms subside in a few hours and the victim falls into a deep sleep and is on his way to recovery.

First Aid and Medical Treatment: Induce vomiting if subject is alert; if sleepy, institute gastric lavage followed by activated charcoal and cathartics. UNDER NO CIRCUMSTANCES GIVE ATROPINE. Otherwise treatment is symptomatic.

PANTHER *AMANITAS* (all parts) potentially dangerous ☠

The panther *Amanita*, *A. pantheria*, and a related variety of mushroom, the booted *Amanita*, *A. cothurnata*, are both hallucinogenic and potentially dangerous mushrooms. The handsome panther has a one- to six-inch-wide convex cap that ranges in color from whitish to brown (brown is more common) and is covered with concentric rings of creamy white patches. As the mushroom ages, the cap flattens and its margins, which have fine radial lines, sometimes change color from yellow-brown to sooty brown. The cap is sticky when wet. The gills are usually free and white in color, and the white stem (two to seven inches long) enlarges at its base to form a scaly bulb. A thin, pliant white skirt or ring encircles the upper part of the stem.

Creamy whitish-brown warts

striate margin

2-inch to 5-inch sticky brownish cap flattens with age

gills are white, crowded

white ring

2½-inch to 5-inch white, hairy stem

collar where bulb and stem meet

Panther Amanita
Amanita pantheria

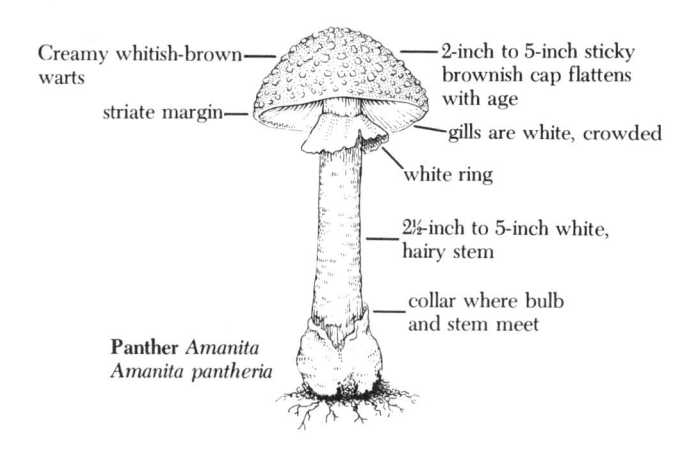

The booted *Amanita* is smaller than *A. pantheria*, having a one- to four-inch-wide white convex cap whose center is yellowish. As it ages, the cap flattens and the fine lines at its margins are seen. The gills are white, free and crowded, and the two- to five-inch-long white stem is encircled by a fine white skirt at its upper end and thickens and becomes scaly at its base. This mushroom is found from midsummer to early fall under hardwoods and conifers in the eastern part of the United States, ranging from Florida in the south to New York state in the north and westward as far as the Pacific Northwest. It is common in the Rocky Mountain states and Pacific Northwest.

Toxins and Symptoms: *Amanita pantheria* and its look-alike, *A. cothurnata*, contain a potent toxin that acts on the central nervous system. This substance, ibotenic acid, is converted in the body to the psychoactive poison muscimol. The amounts of toxin which are concentrated in the tissues under the mushroom's skin vary with the seasons, being highest in the summer months, lowest in the fall. Poisoning with both of these muscimole-containing *Amanitas* causes alternating groups of symptoms. Usually one sees the classic triad of sweating, salivation, and lacrimation, sometimes followed by visual disturbances and relatively mild motor disturbances. In some victims, for reasons unknown, motor symptoms dominate, producing an epileptic-like convulsion. Such symptomology is called toggle-switch intoxication. Though not considered a deadly mushroom, *A. pantheria* has been questionably implicated as a cause of several deaths in the Pacific Northwest.

First Aid and Medical Treatment: Refer to *A. muscaria*, p. 146.

III. MUSHROOMS THAT CAUSE SWEATING, TEARS, AND SALIVATION

CLITOCYBES

Two members of the genus *Clitocybe*, if eaten, produce the so-called muscarine type intoxication. *C. dealbata*, commonly called the sweating mushroom, is a small, dull white mushroom whose smooth, dry, convex cap subsequently flattens with upturned margins to form a shallow, gray-pinkish saucer. It has narrow whitish gills that descend slightly down the one- to two-inch-tall white stem. Its larger cousin, *C. dilatata*, also known as *C. cerussata*, grows in crowded clusters; hence it has been called the crowded white *Clitocybe*. It has a white, smooth, dry cap (two to six inches in diameter) which is initially convex, but then flattens and turns up at the margins which often split. The white to tan gills may attach down the two- to five-inch stem.

Distribution and Habitat: *C. dealbata* is widely distributed, occurring in grassy fields or open woods from July through September (later in California), while *C. dilatata* is found in dense clusters along roadsides in the Pacific Northwest.

Toxins and Symptoms: Both of these *Clitocybes* contain muscarine and produce moderate to severe intoxication dominated by copious sweating, chills, slavering, blurred vision, cramps, pinpoint pupils, watery diarrhea, a slow pulse, and hypotension (low blood pressure). Symptoms normally subside by themselves without treatment six to 24 hours after ingestion. However, people with cardiovascular problems can be killed by muscarine poison-

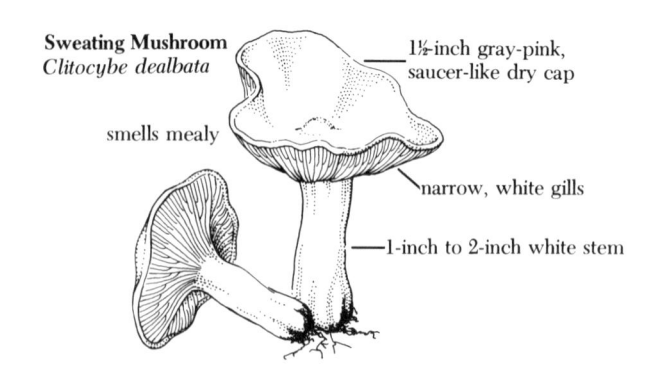

Sweating Mushroom
Clitocybe dealbata

1½-inch gray-pink, saucer-like dry cap

smells mealy

narrow, white gills

1-inch to 2-inch white stem

ing, and deaths in this susceptible group of people have been reported. Severe poisoning will require hospital treatment.

First Aid and Medical Treatment: Empty the stomach and administer 0.5 to 1.0 mg Atropine (repeated in ½ hour if needed). BE SURE that the victim has not also ingested *A. muscaria* or *A. pantheria* since Atropine would exacerbate their toxin. If vomiting has been extensive, restore fluids and electrolyte balance. Atropine usually produces prompt reversal of symptoms.

INOCYBES OR FIBER-HEADS

There are a large number of members of the genus *Inocybe*, commonly called fiber-heads, that are poisonous and should be avoided. Many of these mushrooms emit a distinct odor that varies from strong and fishy to green corn to pungent. Typical of the poisonous *Inocybes* is the straw-colored fiber-head, *I. fastigiata*. This mushroom has a one- to three-inch-wide, yellowish, conical, or bell-shaped cap, with distinct fibers radiating from the darker-colored central knob toward the margin, like spokes on a wheel. As the mushroom ages, the top flattens and the margin turns upward and splits. The narrow white gills are crowded and, though attached to the stem, are nearly free. As the mushroom ages, the gills turn olive-gray. The stem, an inch and a half to three and a quarter inches long, is one-eighth to three-eighths of an inch thick, fibrous and creamy in color. Other poisonous *Inocybes* species are (1) the black-nipple fiber-head, (2) the white fiber-head, (3) the torn fiber-head, (4) the woolly fiber-head, (5) the lilac fiber-head, (6) the blushing fiber-head, (7) the scaly fiber-

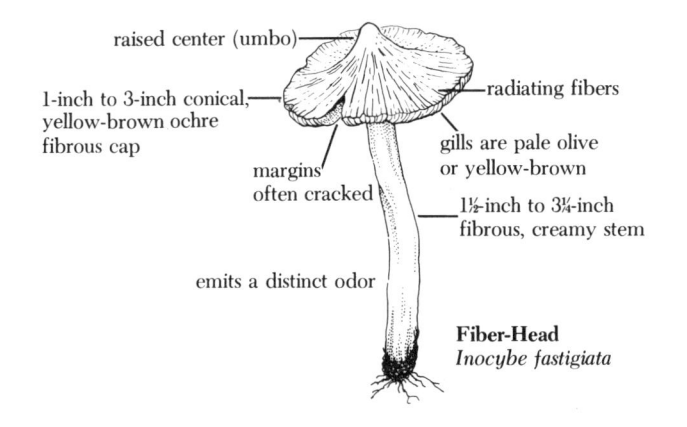

raised center (umbo)
1-inch to 3-inch conical, yellow-brown ochre fibrous cap
radiating fibers
margins often cracked
gills are pale olive or yellow-brown
1½-inch to 3¼-inch fibrous, creamy stem
emits a distinct odor

Fiber-Head
Inocybe fastigiata

150   Poisonous Plants and Mushrooms of North America

head and (8) the pungent fiber-head. We cannot go into detail on all the *Inocybes* in this book, but strongly recommend the Audubon Society's *Field Guide to North American Mushrooms* for field characteristics and distribution.

Symptoms, Toxins, and Treatment: Most *Inocybes* contain muscarine and produce the spectrum of effects described in the section on the Clitocybes. See pages 148 and 149 for symptoms and treatment.

IV. HALLUCINOGENIC MUSHROOMS

THE *PSILOCYBES*

The *Psilocybe* mushrooms have a long history of use in ancient religious ceremonies, being used by the Aztecs of Mexico over 3,000 years ago. They contain at least two non-persistent, potent, mind-altering substances, psilocybin and psilocin, whose effects are very similar to those of lysergic acid diethylamide (LSD). Though about 100 times less potent than LSD, they can produce a spectrum of emotional and psychotic-like symptoms.

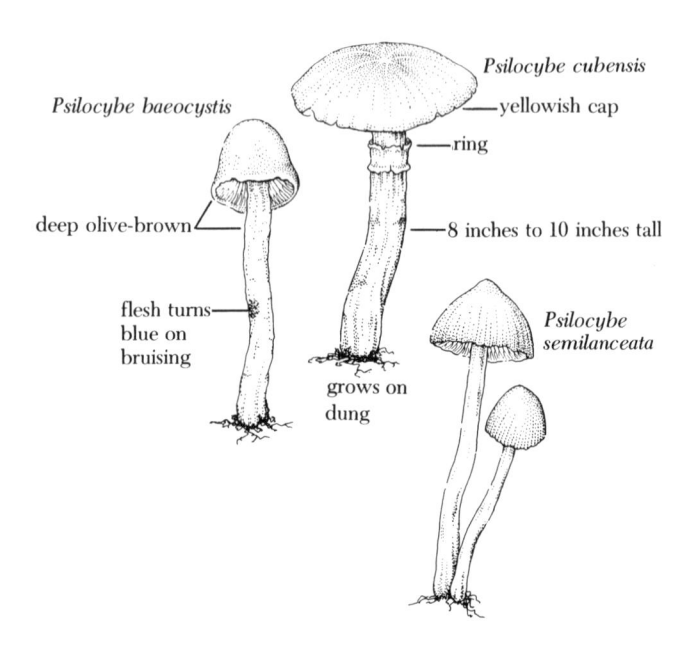

Psilocybe baeocystis

deep olive-brown

flesh turns blue on bruising

Psilocybe cubensis

yellowish cap

ring

8 inches to 10 inches tall

grows on dung

Psilocybe semilanceata

There are many members of the *Psilocybe* genus, and descriptions are inexact for many of them. The most potent are *P. strictipes* and *P. baeocystis*. *P. strictipes* has a yellow or olive-brown cap and stem. The flesh of the stem turns brown when broken. The skirt usually is not present. This mushroom can be found growing under conifers in clusters growing on either bark or mulch, and is frequently seen on bark mulch in moist flower beds. Three or four of these small mushrooms contain enough toxin to produce symptoms. *P. baeocystis* has a deep-olive-brown cap and stem which turns silver-gray as it dries. Its flesh does not discolor brown on bruising, but may show a bluish stain. It can grow on lawns and is quite potent, requiring only two mushrooms to produce hallucinations. There is at least one death of a child attributed to *P. baeocystis*. Another potent *Psilocybe*, which is a native of semitropical America, *P. cubensis*, grows on dung heaps. It is a very large (eight-to-ten-inch tall) mushroom with a convex yellow cap and a conspicuous ring. Only one to three of these king-sized *Psilocybes* will produce intoxication.

Distribution and Habitat: There are several less potent *Psilocybes* which if eaten in large quantities (20 to 40 mushrooms) can produce hallucinations. *P. semilanceata*, which grows in tall grass, and *P. pellicula*, which grows in rotting mulch, both look very similar. The *Psilocybes* are infrequently encountered in the United States and Canada.

Symptoms: Thirty to sixty minutes after eating a sufficient number of *Psilocybes*, there is a progressive loss of concentration and understanding. Some people get anxious, and there are changes in sensory perception, such as an acute sensitivity to touch. There are often kaleidoscopic visions with size, color, depth of vision, and shape being distorted, as was described by the Beatles in their song, "Lucy in the Sky with Diamonds." The visual hallucinations with altered colors and light are particularly acute with the eyes closed. Some people may experience a panic reaction and mood swings are usually toward euphoria but some people experience depression. Children seem to be much more sensitive than adults and may show a fever or even convulse; at least one death has been recorded. The symptoms may last several hours, and in severe poisonings the victim usually recovers without lasting effects in five to ten hours. A more detailed personal account of *Psilocybe* effects can be found in Haard's marvelous little book, *Poisonous and Hallucinogenic Mushrooms*.

First Aid and Medical Treatment: To get the poison source out of the system, induce vomiting and give cathartics. Recovery is usually spontaneous.

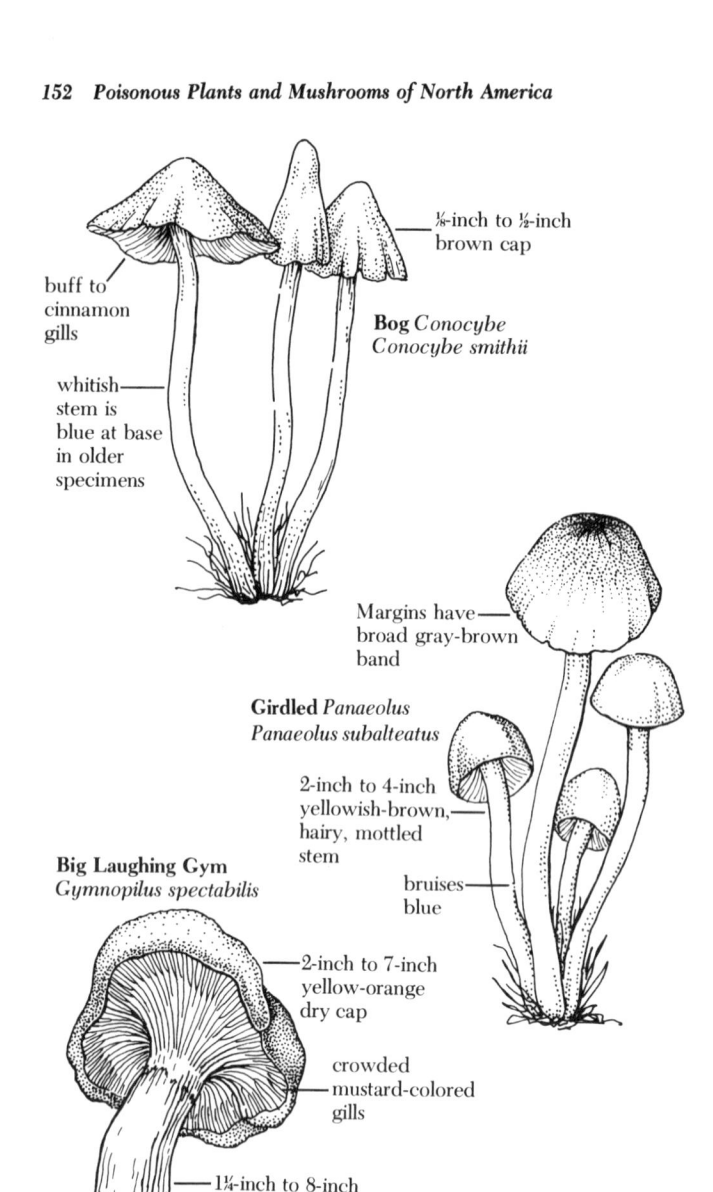

152 Poisonous Plants and Mushrooms of North America

⅛-inch to ½-inch brown cap

buff to cinnamon gills

Bog *Conocybe*
Conocybe smithii

whitish stem is blue at base in older specimens

Margins have broad gray-brown band

Girdled *Panaeolus*
Panaeolus subalteatus

2-inch to 4-inch yellowish-brown, hairy, mottled stem

bruises blue

Big Laughing Gym
Gymnopilus spectabilis

2-inch to 7-inch yellow-orange dry cap

crowded mustard-colored gills

1¼-inch to 8-inch club-shaped, yellow-orange stem

bitter, licorice odor

THE BOG *CONOCYBE*

Widely scattered in the moss bogs of the Pacific Northwest is a small brown-capped mushroom, *Conocybe smithii*, which contains varying concentrations of hallucinogenic secondary products. This bog *Conocybe* has a small conical brown cap (one-eighth to half an inch wide), which flattens with a central knob. Its narrow, buff to cinnamon-brown gills attach to a thick, whitish stem that turns blue at the base as the mushroom matures.

Distribution and Habitat: In moss bogs of the Pacific Northwest and north central states from September to November.

Toxins and Symptoms: Similar to *Psilocybes*, see p. 151.

First Aid and Medical Treatment: Similar to *Psilocybes*, see p. 151.

THE GIRDLED *PANAEOLUS*

P. subalteatus is a common, widely distributed hallucinogenic mushroom with a conical, knobbed, smooth, moist, tan-tawny brown cap whose margins have a broad, darker brown-gray band. Its close brownish gills, which become darkly mottled with age, are attached to a two- to four-inch-long hairy, reddish-brown, mottled stem which may bruise blue.

Distribution and Habitat: These mushrooms grow in groups or as scattered individuals in dung heaps or richly manured soil across much of the United States and Canada. They are early risers, appearing in June and July.

Toxins, Symptoms, and Medical Treatment: See *Psilocybes*, p. 151.

GYMNOPILUS SPECTABILIS

Commonly called big laughing gym, *G. spectabilis* is a large, wood-dwelling, yellow-orange hallucinogenic mushroom. Its large convex, ochre-orange cap (up to seven inches across) flattens with a slight central knob as it matures. It may have tiny scales, emits a slight anise or licorice odor and tastes very bitter. It has crowded yellowish-brown gills attached to a thick, sometimes up to eight-inch-tall, yellow-ochre, slightly fibrous stem, around which may be a ring.

Distribution and Habitat: Common and widely distributed in North America, where it grows in clusters on stumps or over decaying deciduous and conifer wood.

Toxins and Symptoms: As its name, big laughing gym, implies, it contains hallucinogens which in some people produce euphoria and gales of laughter. Its as yet uncharacterized toxins are possibly related to those of the *Psilocybes*. See p. 151 for symptoms in more detail and treatment.

V. MUSHROOMS TOXIC ONLY WITH ALCOHOL CONSUMPTION

INKY CAPS

Coprinus atromentarius is a most unusual intoxicating mushroom that is common in many parts of the United States. Other common members of the genus *Coprinus* (the name comes from the Greek word for dung) are edible, e.g., *C. comatus* and *C. micaceus*, but *C. atromentarius* contains a substance called coprine that can make you violently ill if you drink any alcohol-containing beverage. Coprine acts specifically to inhibit the liver enzyme aldehyde dehydrogenase, which normally metabolizes ethyl alcohol. One product of alcohol breakdown is a substance called acetaldehyde,

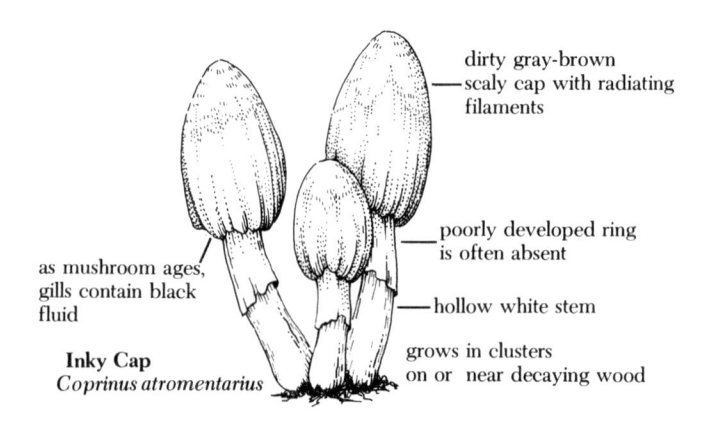

Inky Cap
Coprinus atromentarius

- dirty gray-brown scaly cap with radiating filaments
- poorly developed ring is often absent
- hollow white stem
- grows in clusters on or near decaying wood
- as mushroom ages, gills contain black fluid

but this is normally broken down into harmless by-products rather rapidly. In the presence of coprine, acetaldehyde accumulates and makes the victim violently sick (see symptoms below). One trial treatment for chronic alcoholics was a drug called Antabuse®, whose action is similar to that of coprine. The rationale for giving this drug was the compulsive drinker would not voluntarily drink alcohol if he knew he would become ill.

Description, Distribution, and Habitat: This genus is usually called "inky caps" because as the mushrooms age their gills gradually break down and accumulate an inky-black fluid. *C. atromentarius* has a smooth conical to bell-shaped cap that is a sort of dirty grayish-brown, sometimes scaly toward the middle. The cap has thin, more or less parallel filaments, with radiating furrows or lines at its margin. The hollow white stem tapers slightly upward from a wider base. The skirt is not very well developed and in mature mushrooms of this species often disappears. Mushrooms of this genus are widely distributed across Canada and the United States. They are usually found in clusters on the ground on or near decaying wood, or near the base of trees, but can also be found in fields or gardens. In warmer climates they may be found from November to April.

Symptoms: Peak effect of coprine is reached six hours after consumption and may last for over 48 hours before the normal enzyme levels are restored. With alcohol consumption the symptoms include marked flushing of the skin, lowered blood pressure, accelerated heart rate and increased rate of respiration. Often there is a metallic taste in the mouth, numbness of the extremities, and palpitations of the heart. Nausea and vomiting usually occur.

Prevention, First Aid, and Medical Treatment: The best advice is: learn to recognize the mushroom. When in doubt, stay away from drinking any alcoholic beverage for two days. Once symptoms occur, they will run their course over several hours and then recovery is complete and spontaneous.

VI. MUSHROOMS THAT CAUSE GASTROENTERITIS

There is a very large and heterogeneous group of mushrooms which contain some as yet undetermined toxins that cause mild to severe gastroenteritis that lasts from two hours to two days.

Some only produce minor upsets and are not always toxic, for even within the same species toxicity varies with both locale and season. There is also some individual variation in the responses of different people and a mushroom that has no effect on one person can make another violently ill. It is impossible to cover all of the mushrooms that cause nausea, vomiting, and diarrhea, thus in the following pages we have only included some of the major villains of the bowel.

RUSSULA EMETICA

A common, conspicuous, boldly colored summer mushroom that can produce very severe gastrointestinal symptoms is *Russula emetica*. Fortunately, it has a couple of distinguishing characteristics that make certain recognition easy. First, the cap is fragile and if thrown against a tree will shatter. Secondly, a small fragment, if tasted at the tip of the tongue, will give an acrid or peppery taste. Eaten raw it can cause severe emesis (vomiting), hence the name *emetica*. Supposedly harmless if cooked, it is still best to avoid this species.

Description: *R. emetica* has a very distinctive pure red or scarlet cap (two to four inches broad) with pink-red flesh and white gills. The cap is usually convex, but as the mushroom ages and spreads, the cap flattens with a depressed center. The margins of the cap are striated. In older specimens exposed to sunlight the scarlet cap may fade to a light pink. The white stem is usually longer than the cap is wide.

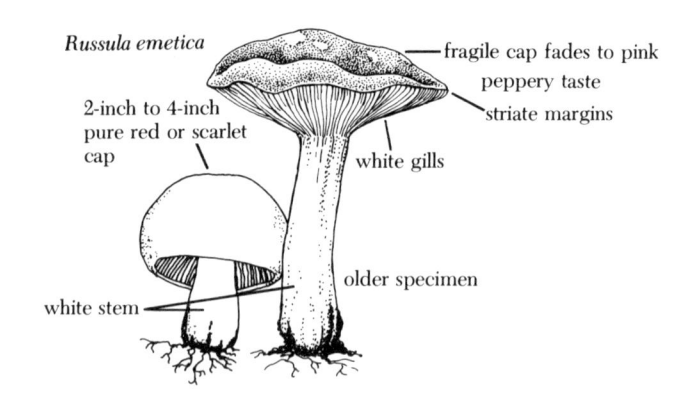

Distribution and Habitat: Usually found in late spring through the summer and even early fall if there is sufficient moisture. Most often found under conifers. They are widespread throughout Canada and the United States.

Symptoms: Within a few hours after eating raw *R. emetica*, nausea and vomiting begin. The vomiting may be severe and may be accompanied by intensely painful abdominal cramps. Diarrhea may follow the vomiting as the irritant proceeds down to the lower bowel. Symptoms taper off and end spontaneously after a few hours of unpleasantness.

First Aid and Medical Treatment: Have the victim rest, restore fluids slowly, provide a light diet. Anti-emetics may help once the stomach contents are emptied.

Milky Mushrooms

The *Lactarius* group of mushrooms contain a milky latex which oozes from any cut or bruised part of the fruiting body. While some members of the genus are edible, the toxic members generally have a sharp, acrid taste and exude white latex that turns yellow upon exposure to air. Thus, it would be wise to avoid all *Lactarius* whose milk changes to yellow, yellow-green or pale purple. Among the *Lactarius* that are known and thought to be toxic are the northern bearded milky (*L. repraesentaneus*), a large, yellow-orange capped (up to eight-inches wide), widely distributed mushroom whose latex turns purple-violet when exposed to air. *L. rufus* is a bay-red capped inhabitant of pine forests

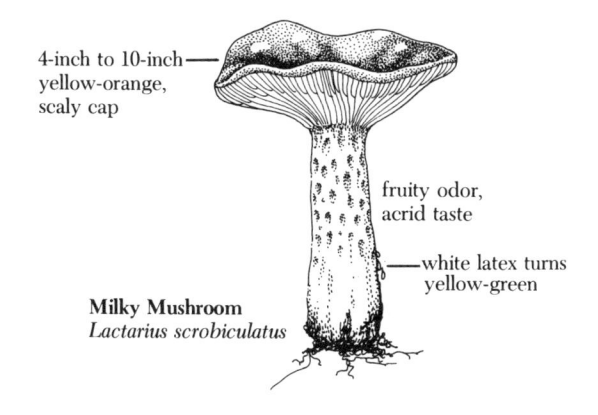

4-inch to 10-inch yellow-orange, scaly cap

fruity odor, acrid taste

white latex turns yellow-green

Milky Mushroom
Lactarius scrobiculatus

and sphagnum bogs that is widely distributed from August to November. *L. scrobiculatus* is a large, yellow-orange capped (up to ten-inches broad) mushroom whose scaly domed cap becomes funnel-shaped as the margins roll up. Its odor is fruity but its taste is burning and peppery.

Toxins and Symptoms: The latex contains a potent gastrointestinal irritant; hence the symptoms would be that of any plant or fungal irritant. Symptoms begin within 30 minutes to three hours after ingestion, and full recovery may take a day or two.

Treatment: Symptomatic—treat as you would any gastroenteritis.

JACK O'LANTERN OR FALSE CHANTERELLE

The jack o'lantern of the genus *Omphalotus* is a startling looking bright orange or yellow-orange mushroom that grows in clusters on wood or decaying buried wood. In the dark its gills produce an eerie green phosphorescent glow. Its three- to eight-inch-wide convex cap flattens as it matures, forming a shallow depressed cup as the margins turn upward. A small knob is often found in the center of the cup. At this stage the cap is dry, smooth, and yellowish. The narrow orange-yellow phosphorescent gills continue down the orange stem, which is from three to eight inches long and tapers, narrowing and darkening toward the base. (**Plate 32.**)

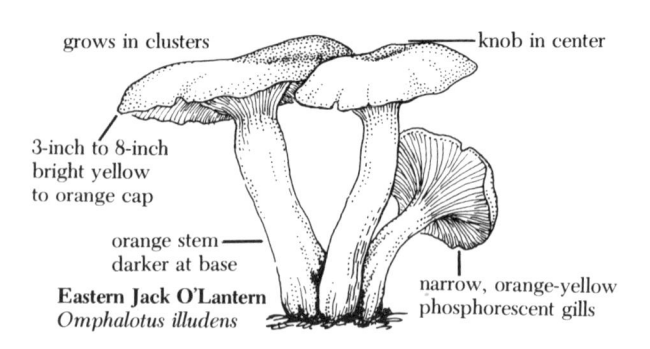

Distribution and Habitat: The eastern jack o'lantern (*Omphalotus illudens*, also known as *Clitocybe illudens*) and the western jack o'lantern, *O. olicascens*, are found in clusters in deciduous woods, often associated with oak trees. In eastern North America they can be found from July to November, while in the West they are found from November to March. They are widely distributed all across the United States.

Toxins and Symptoms: The jack o'lanterns contain a gastrointestinal irritant that produces nausea, vomiting, stomach cramps and diarrhea which may last only a few hours or may produce symptoms for two days.

First Aid and Medical Treatment: Treat symptomatically and make sure to prevent dehydration and electrolyte imbalance if vomiting and diarrhea is extensive.

HEBELOMAS

The poison pie, *Hebeloma crustuliniforme,* is a widely distributed, though rather uncommon, mushroom that emits a radish-like aroma and can cause some unpleasant gastrointestinal symptoms if eaten. As a matter of fact, *all members* of the genus *Hebeloma* should not be eaten. The poison pie has a creamy convex cap one and a quarter to three and a half inches across which flattens as it matures. The center of the cap is light reddish-brown, and the whole cap is slimy-sticky to the touch. Its crowded, narrow, creamy to buff gray-brown edged gills are often covered with water droplets and become mottled brown when dry. The white stem is thick (up to two and a quarter inches in diameter), and is generally hairy.

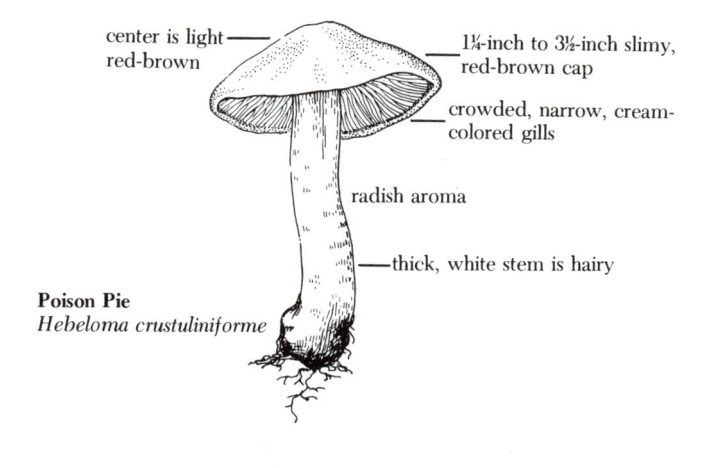

center is light red-brown

1¼-inch to 3½-inch slimy, red-brown cap

crowded, narrow, cream-colored gills

radish aroma

thick, white stem is hairy

Poison Pie
Hebeloma crustuliniforme

Distribution and Habitat: This mushroom is often found in western forests, in both conifer and deciduous woods, but may also grow in residential urban environs. It may occur as a few scattered individuals, in groups or in fairy rings. In cooler areas, it is usually found from September through November, but in California's more balmy clime may last through May.

Toxins and Symptoms: A variety of gastrointestinal irritants that will cause nausea, vomiting, malaise, cramps, and diarrhea within 30 minutes to three hours after ingestion. Symptoms may persist for a few hours or may last up to two days, depending upon the amount of toxin ingested and the individual victim's state of health.

Treatment: Symptomatic.

ENTOLOMA

There are at least three toxic members of the genus *Entoloma* that can cause mild to severe gastroenteritis. These are: *E. sinuatum*, the lead poisoner, *E. strictius* and *E. vernum*, the spring *Entoloma*. *E. sinuatum* emits a strong cucumber-like odor and initially has a dirty brown-gray, slippery convex cap (three to six inches in diameter) which flattens with age. Its margins are rolled inward. The pale yellow-gray gills turn pink in older mushrooms and attach to a thin (up to one-inch-long) stem that is hollow and pale gray. This common, widely distributed mushroom can be

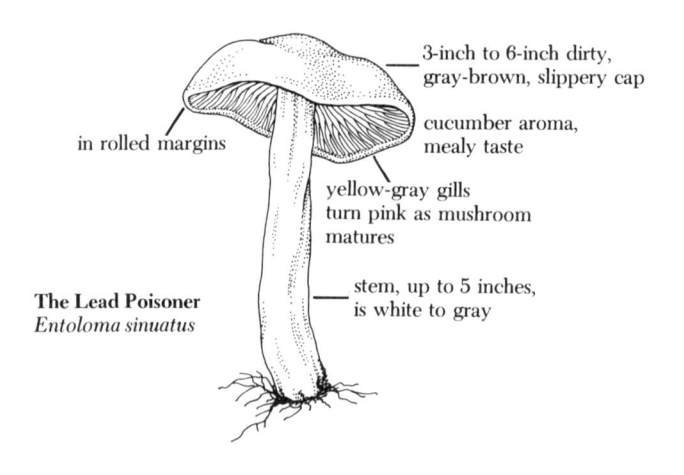

Illustration labels:
- 3-inch to 6-inch dirty, gray-brown, slippery cap
- in rolled margins
- cucumber aroma, mealy taste
- yellow-gray gills turn pink as mushroom matures
- stem, up to 5 inches, is white to gray

The Lead Poisoner
Entoloma sinuatus

found August through September in groups or scattered under hardwoods and conifers.

E. strictius is smaller, having a conical or bell-shaped gray-brown knobbed cap one to two inches in diameter that flattens with age. The white gills are close, turn pink in older specimens and attach to a two- to four-inch-long gray-white stem whose base is covered with white threads. It can be found from April to September in wet areas of woodlands from Canada to Florida east of the Great Lakes.

E. vernum is another small mushroom with a distinctly knobbed, bell-shaped light brown cap which becomes convex later. The gills, buff to pinkish brown, attach to a brownish stem one to four inches long with white threads at its base. This early spring mushroom (April to June) is frequently found along paths in woods from Canada, New York, and New England to Wisconsin.

Toxins: Not really known, but they do affect the gastrointestinal system.

Treatment: Symptomatic.

VERPA BOHEMICA (CHOLERA-LIKE SYMPTOMS IN SENSITIVE INDIVIDUALS)

This species is another morel look-alike that is considered a prized edible species by some gourmets. However, some people are sensitive to it on first tasting; others, after years of eating this mushroom, develop sensitivity and show symptoms. While not considered particularly poisonous, caution is suggested.

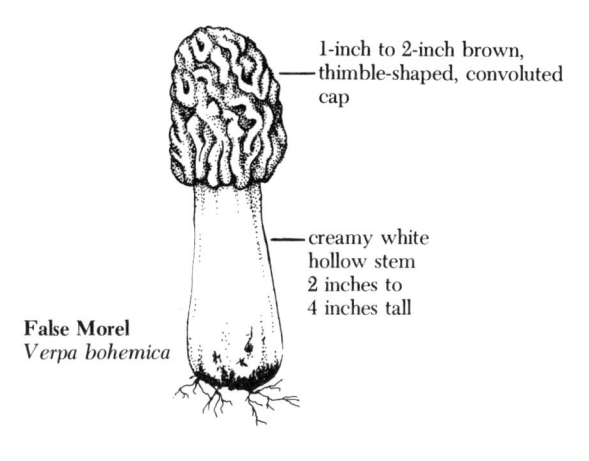

1-inch to 2-inch brown, thimble-shaped, convoluted cap

creamy white hollow stem 2 inches to 4 inches tall

False Morel
Verpa bohemica

Description: *V. bohemica* is two to four inches tall and has a brownish, thimble-shaped cap (one to two inches across) that is vertically furrowed and wrinkled and hangs down from the apex of the stem which is a long, creamy, white, hollow cylinder.

Distribution and Habitat: *Verpa* can be found in moist woodlands, frequently associated with cottonwood trees, from March to early June in cooler climates.

Symptoms: Gastrointestinal upsets similar to bacterial or viral gastroenteritis. They are usually temporary, ranging from mild to severe. Starting a few hours after ingestion, the victim will experience queasiness, nausea, and cramps, and possibly vomiting and diarrhea. Recovery occurs within a few hours and is spontaneous.

First Aid and Medical Treatment: If the severe response occurs, it will usually be accompanied by vomiting; hence gastric lavage is not called for. Treat symptoms, and recommend light foods and rest.

POISONOUS TUBE FUNGI, THE *BOLETES*

While most mushrooms have gills, *Boletes* species have a collection of spore-bearing tubes under their caps which give the undersides a porous or spongy look, and the spores either drop out or are forcibly expelled. There are four poisonous *Boletes* species which contain a still obscure irritant toxin that causes gastrointestinal upsets. Fortunately the poisonous types have some good field identification characteristics. Never eat a *Bolete* whose tube mouths are red or whose flesh turns blue or blue-green when cut or bruised!

Boletus eastwoodiae is a large lovely *Bolete* with a convex olive to dark brown cap (four to nine inches wide), brilliant scarlet tube mouths, and a thick, three-to eight-inch tall, club-shaped stem that is heavily ridged and veined. Its flesh quickly turns blue if bruised. A common mushroom, it can be found in the fall rainy season under conifers on the west coast.

Boletus satanus, an uncommon *Bolete*, has an off-white to olive convex cap (four to nine and one-half inches broad) attached to a blood red bulbous stem (two to seven inches tall) which is pink over the bulb. Tube mouths are blood-red. The flesh turns blue if cut or bruised. Sometimes found in the southern United States and under conifers on the west coast.

Poisonous Mushrooms

Poisonous Tube Fungus
Boletus eastwoodiae

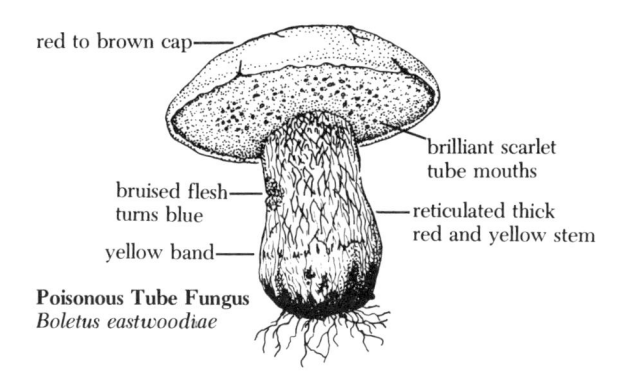

- red to brown cap
- brilliant scarlet tube mouths
- bruised flesh turns blue
- reticulated thick red and yellow stem
- yellow band

Boletus subvelotipes has a velvety, bright ochre convex cap that turns brown in older specimens. Its stalk is covered with minute brown particles. The tube mouths are red and tissues turn blue quickly if damaged. They are found under hardwoods in the summer in the Great Lakes Region.

Boletus miniato-divaceus is a large, brilliantly colored *Bolete*. Its brick-red (four to nine and one-half inch wide) convex cap bleaches out to a dull orange as it ages. The tubes are a brilliant chrome-yellow. The flesh is yellow and it quickly stains blue or blue-green if damaged. The club-like stem is yellow at its top, grading into red to orange at the base. It occurs east of the Great Plains in mixed woods during the summer months in the south and in early fall in the north.

Boletus luridus has a convex, moist or viscid, olive, yellow, red, or brown cap, red tube mouths, a club-shaped stem which is yellow-orange near the cap grading into red at the base. It is found in central and eastern United States in the summer and fall under mixed woods. A similar species, *B. calupus*, is found on the west coast under conifers from late summer to fall.

Symptoms: Thirty to 120 minutes after ingestion a variety of gastrointestinal symptoms occur although there seems to be considerable variability in severity of the symptoms. These include nausea, vomiting, painful abdominal cramps, and often diarrhea. The symptoms usually taper off after a few hours and normal bowel function returns in a day or so.

First Aid and Medical Treatment: If vomiting or diarrhea has been severe, watch for electrolyte or water imbalance. Treat the symptoms conservatively, e.g., recommend light food, slow fluid replacement, and rest.

PAXILLUS INVOLUTUS

This *Paxillus* species, although considered edible in some parts of the country, may produce a gradual immune hypersensitivity in some individuals. The mushroom is also palatable in some areas and sour-tasting in others. *P. involutus* has a convex brown cap two to five and three-quarters of an inch in diameter that flattens and then becomes saucer-like. It is slimy or sticky when moist and covered with matted fibers. The cap margins are in-rolled, and yellow-olive gills descend down the yellow-brown stem, two to four inches long, that is often stained brown. The gills, if bruised, also turn brown and are easily separated from the cap.

Distribution and Habitat: Widely found near or on wood, usually associated with conifers, across much of North America.

Toxins and Symptoms: Symptoms, if they occur, will develop in an hour to several hours after ingestion and will occur only in people who have eaten the poisonous pax before. Nausea, vomiting, diarrhea, and altered cardiovascular function may be present. There may be a breakdown of red cells and subsequent kidney involvement. In these sensitized individuals, symptoms usually last only a few days, but if the amount ingested is large and the individual is very sensitive, the symptoms could be prolonged and serious enough to require hospitalization.

First Aid and Medical Treatment: Treat symptomatically as you would for any food allergy.

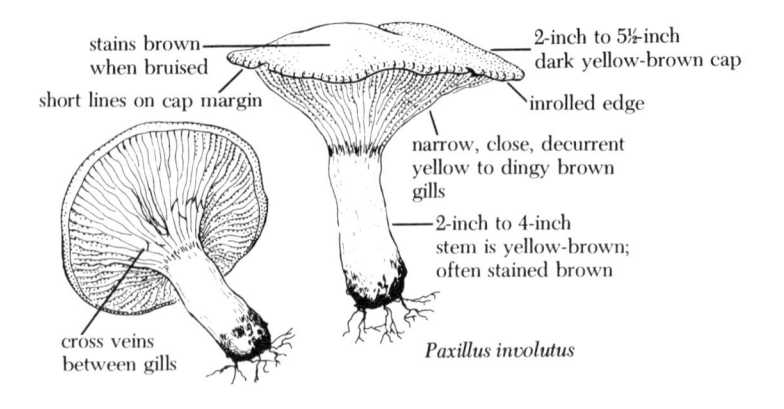

Paxillus involutus

- stains brown when bruised
- short lines on cap margin
- 2-inch to 5½-inch dark yellow-brown cap
- inrolled edge
- narrow, close, decurrent yellow to dingy brown gills
- 2-inch to 4-inch stem is yellow-brown; often stained brown
- cross veins between gills

CORTINARIUS (kidney toxin) ☠

If there ever was a biological misnomer, *Cortinarius gentilis*, the gentle cort, would certainly win first prize because, despite its name, it has been reported to contain a deadly delayed kidney poison, but surprisingly, no cases of poisoning have been reported yet. *C. gentilis* has a one- to two-inch-wide russet brown conical cap, which becomes bell-shaped and rusty yellow-brown on maturing. Its broad, yellow-brown to rusty gills attach to a short, thick, (one- to three-inch-long), yellow to brownish-orange stem.

Distribution and Habitat: This little brown mushroom is common and widespread under conifers in northern portions of North America in September and October. In the Rockies it occurs earlier. Two European species of mushrooms also contain this same delayed kidney poison. They are *Cortinarius orellanus* and *Lepiota helveola*. Neither have been reported in North America.

OTHER POISONOUS *AMANITAS*

There are a number of *Amanitas* that are reported to be poisonous or suspected of being poisonous. These mushrooms may contain a number of toxins which, among other things, will cause nausea, vomiting, and diarrhea, but may have other effects. Several of these are included in the text. Given the insidious nature of this genus, it would be wise to avoid eating any *Amanita*, since even those species that some authors list as possibly edible may hybridize with toxic species and be toxic.

A. flavorubescens and *A. flavonia* are possible poisonous look-alikes. *A. flavorubescens* has a light yellow cap (two and one-half to five inches broad) with a slightly olive tinge, sometimes bearing scattered pale yellow warts. In young specimens, the cap is roundish (ovoid). The skirt-like ring is fragile and the cup adheres tightly to the firm fleshy stem. The gills are white to gray or may even be yellowish. The cap separates easily from the stem and the cup and other parts stain red on bruising. They are infrequently found on grassy edges of hardwoods but sometimes are found on lawns or meadows in the Southeast, Northeast, and central southern United States during the summer and fall. *A. flavoconia* is more frequent and is similar in appearance, but its cap is somewhat smaller and more yellow to orange and has pale yellow warts. The cup is also yellow.

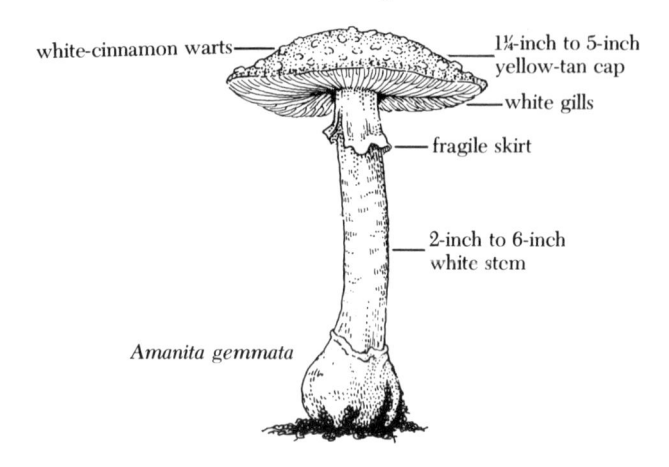

166 Poisonous Plants and Mushrooms of North America

white-cinnamon warts — 1¼-inch to 5-inch yellow-tan cap
— white gills
— fragile skirt
— 2-inch to 6-inch white stem

Amanita gemmata

A. *chlorinosoma*, probably poisonous, is a robust stately mushroom with a powdery white cap (two and one-half to ten inches wide) adorned by small cottony warts in the center. Its creamy white gills are crowded and its white stalk (four to ten inches tall) is encircled by a fragile skirt near the cap. The volva remains only as patches on the widened bulbous end of the stem. It is clearly identifiable because it smells like chlorine. Common under hardwoods in the southeastern United States. The toxin is yet to be discovered.

A. *citrina*, another possibly poisonous member of this genus, has a convex green-yellow cap (one and one-half to four inches broad) sometimes with adhering pinkish patches and a smooth whitish in-curved margin. Its gills are white and the stem (three to five inches tall) is girdled by a white skirt-like ring just under the cap that terminates in a flattened bulb with a tight-fitting volva which may form a collar-like ring at the base of the stem. Found frequently in the East, Southeast and South under pines or hardwoods from summer through late fall. The toxin in this species may be bufotenine, which has central nervous sytem effects.

A. *gemmata*, which is listed only as suspect in some mushroom books, can be quite poisonous. Dr. Lot B. Page of Newton-Wellesley Hospital in Newton, Massachusetts, reported a confirmed case of mushroom intoxication caused by this species and recently two more cases of A. *gemmata* intoxication have been reported in New England. A. *gemmata* is a common mushroom with a broad yellow-tan convex cap (one and a quarter to five inches broad) which flattens with age. White to cinnamon-colored warts cover the cap but they may wash off in heavy rains. The cap margin has fine vertical lines and the gills are white. The stem, two to six inches long, is white and encircled with a fragile skirt. The cup, or volva, adheres tightly to the bulbous base of the stem, forming ragged concentric rings. Widely distributed in mixed open woods from late spring to fall. The toxin is unknown and may affect the heart muscle.

SELECTED REFERENCES

ADVENIER, C., B. MALLARD, M. SANTAIS & F. RUFF. 1982. The effects of atropine on anaphylactic shock in the guinea pig. Agents and Actions 12:103-107.

ANONYMOUS. 1981. Sunscreens, photo carcinogenesis, melanogenesis, and psoralens. British Medical Journal 283:335-336.

ARENA, J.M. 1979. *Poisoning; Toxicology, Symptoms, Treatment.* Charles C. Thomas, Springfield.

BECKER, C., T. TONG, F. BARTTER, U. BOERNER, R. ROE, R. SCOTT, M.B. MACQUARRIE & F. BARTTER. 1976. Diagnosis and treatment of *Amanita phalloides*-type mushroom poisoning. Western J. Med. 125:100-109.

BERGERS, W. & M.A. GERRIT. 1980. Toxic effect of the glycoalkaloids solanine and tomatine on cultured neonatal rat heart cells. Toxicology Letters 6:29-32.

BLAW, M.E., M.A. ADKISSON, D. LEVIN, J.C. GARRIOTT & R.S. TINDALL. 1979. Poisoning with Carolina Jessamine (*Gelsemium sempervirens* [L.]) Ait. Journal of Pediatrics 94:998-1001.

BURKE, M.J., D. SIEGEL & B. DAVIDOW. 1979. Anaphylaxis: Consequences of yew (*Taxus*) needle ingestion. N.Y. State Journal of Medicine 79:1576-1577.

CARLTON, M.D., E. BRUCE, E. TUFTS & D.E. GIRARD. 1979. Water hemlock poisoning complicated by Rhabdomyolysis and renal failure. Clinical Toxicology 14:87-92.

CHILTON, W.S. & J. OTT. 1976. Toxic metabolites of *A. pantherina, A. cothurnata, A. muscaria* and other *Amanita* species. Lloydia 39:150-157.

DUNN, I.S., D.J. LIBERATO, R. DENNICK, N. CASTAGNOLI & V.S. BYERS. 1982. A murine model system for contact sensitization to poison oak or ivy urishiol components. Cell. Immun. 68:377-388.

EASON, J.M. AND F.H. LOVEJOY, JR. 1979. Efficacy and safety of gastrointestinal decontamination in the treatment of oral poisoning. Ped. Clin. N. Am. 26:827-836.

EASON, J.M. AND F.H. LOVEJOY, JR. 1983. Narcotic poisoning. *In:* Haddad & Winchester (eds.) *Clinical Approach to Poisons and Drug Overdose.* W.B. Saunders, Philadelphia.

EPSTEIN, W. AND V. BYERS. 1977. The immunology of allergic contact dermatitis. Drug Therapy 7:112-122.

GELLIN, G.A., R. WOLF & T. MILBY. 1971. Poison ivy, poison oak, and poison sumac — Common causes of occupational dermatitis. Archives Environmental Health 22:280-286.

HANSTEEN, V., D. JACOBSEN, K. KNUDSEN, A. REIKVAM, & B. SKUTERUD. 1981. Acute, massive poisoning with Digitoxin: Report of seven cases and discussion of treatment. Clinical Toxicology 18:679-692.

HARDIN, J. & J.M. ARENA. 1974. *Human poisoning from native and cultivated plants.* Duke University Press, Durham.

HARVEY, J. & D. COLIN-JONES. 1980. Mistletoe hepatitis. British Medical Journal 282:186-187.

HEDAYAT, S., D.D. FARHUD, K. MONTAZAMI & P. GHADIRIAN. 1981. The pattern of bean consumption; Laboratory findings in patients with favism and G-6-P-D deficient and a central group. Journal of Tropical Pediatrics 27:110-113.

IVANKOVICH, A.D., B. BRAVERMAN, R.P. KANURU, H.J. HEYMAN & R. PAULISSIAN. 1980. Cyanide antidotes and methods of their administration in dogs: A comparative study. Anesthesiology 52:210-216.

IVIE, G.W., D.L. HOLT & M.C. IVEY. 1981. Natural toxicants in human foods: Psoralens in raw and cooked parsnip root. Science 213:909-910.

KINGSBURY, J.M. 1964. *Poisonous plants of the United States and Canada.* Prentice-Hall, Englewood Cliffs.

KINGSHORN, D.A. (ed.). 1979. *Toxic Plants.* Columbia University Press, N.Y.

KUPCHAN, S.M. AND R.L. BAXTER. 1975. Mezerein: Antileukemic principle isolated from *Daphne mezereum* L. Science 187:652-653.

LEWIS, W.H. & M. ELVIN-LEWIS. 1977. *Medical Botany: Plants Affecting Man's Health.* John Wiley and Sons, N.Y.

LINCOFF, G. and D.H. MITCHEL. 1977. Toxic and hallucinogenic mushroom poisoning. *In:* W.K. Williams (ed.) *A Handbook for Physicians and Mushroom Hunters.* Van Nostrand Reinhold Co., N.Y.

MCGUINAN, M.A. AND F.H. LOVEJOY, JR. In press. The Acute poisoning — Diagnosis and Management. *In: The Emergency.* H. May (ed.). Harvard University Press, Cambridge.

MCHENRY, L.E. & P. BLANK. 1978. Adverse reactions to plants in Florida. Journal of the Florida Medical Association. Volume 65. (Numerous articles in this volume are devoted to poisonous plants.)

MCMILLAN, M. & J.C. THOMPSON. 1979. An outbreak of suspected solanine poisoning in schoolboys: Examination of criteria of solanine poisoning. Quarterly Journal of Medicine 48:227-243.

MITCHELL, J. & A. ROOK. 1979. *Botanical Dermatology: Plants and Plant Products Injurious to the Skin.* Greengrass, Vancouver.

MOHER, L.M. & S.A. MAURER. 1979. *Podphyllum* toxicity: Case report and literature review. Journal of Family Practice 9(2):237-240.

MUENSCHER, W.C. 1939. *Poisonous Plants of the United States.* MacMillan, N.Y.

RUMACK, B.H. AND E. SALZMAN (eds.). 1978. *Mushroom Poisoning: Diagnosis and Treatment.* C.R.C. Press, West Palm Beach.

SCHULTES, R.E. 1979. Hallucinogenic plants: Their earliest botanical descriptions. Journal of Psychedelic Drugs 11:25-28.

SHAW, D. & J. PEARN. 1979. Oleander poisoning. Medical Journal of Australia 2:267-269.

SHERVETTE, R.E., M. SCHYDLOWER, R.M. LAMPE & R.G. FEARNOW. 1979. Jimson "Loco" weed abuse in adolescents. Pediatrics 63:520-523.

SLATER, G.E., B.H. RUMACK & R.G. PETERSON. 1978. *Podophyllum* poisoning: Systematic toxicity following cutaneous application. Obstetrics and Gynecology 52:94-96.

STEVENS, R.H. 1981. Immunoglobulin-bearing cells are a target for the antigen-induced inhibition of pokeweed mitogen-stimulated antibody production. Journal of Immunology 127:968-972.

STIRPE, F., R. LEGG, L. ONYON, P. ZISKA & H. FRANZ. 1980. Inhibition of protein synthesis by a toxic lectin from *Viscum album* L. (mistletoe). Biomedical Journal 190:843-845.

STOUT, G.H., W.G. BALKENHOL, M. POLING & G.L. HICKERNELL. 1970. The isolation and structure of *Daphne* toxin, the poisonous principle of *Daphne* species. Journal of the American Chemical Society 92:1070-1071.

SWAIN, T. (ed.). 1972. *Plants in Development of Modern Medicine*. Harvard University Press, Cambridge.

INDEX

Abrin, 58
Acetylcholine, 26
Aconitine, 120
Activated charcoal, 6
Aesculin, 76
Ajacine, 120
Alder buckthorn, 117–118
Alfalfa, 56
Alkaloids, 6, 31, 34, 36, 37, 38, 41, 42, 46, 49, 50, 56, 62, 75, 92, 100, 104, 107, 108, 112, 113, 120, 122, 124, 125
Alkamines, 66
Allergic contact dermatitis, 12
Amaryllis, 4
Amatoxins, 138
American bittersweet, 114
American ginseng, 74
American holly tree, 124–125
Amines, 121
Amygdalin, 91
Anacardiaceae, 9
Anacardic acid, 23
Anaphylactic shock, 19
Andromedotoxin, 87
Angel's trumpet, 39–40
Angelica, 102
Anise, 102
Apocynaceae, 83, 108
Apple of Sodom, 66
Apple tree, 90–91
Araceae, 68
Araliaceae, 74
Arbutin, 87
Aroid family, 68–71
Arrow-arum, 68
Arrow-wood, 81
Asian bittersweet, 114
Asparagus, 46, 51
Asters, 12
Atamasco lily, 49
Atropine, 7, 38, 43, 87, 120, 146, 149
Atropine sulfate, 107
Autumn crocus, 46–47, 50–51, pl. 8
Azaleas, 86
Aztecs, 30, 35

Baneberry, 110
Barbados nut, 79
Barbiturates, 39, 104
Be-Still tree, 85
Bean family, 56–62
Bear grass, 46
Bell pepper, 62
Bellyache bush, 79
Benzodiazipines, 32, 37, 39, 125
Beta-phenylethylamine, 121
Big laughing gym, 152–154
Bittersweet family, 114–115, pl. 25

Black cherry, 90
Black haw, 81
Black henbane, 44
Black locust tree, 56, 59–60
Black nightshade, 64–65
Black snakeroot, 46, 106
Bladder pod, 56
Blood lily, 49
Bloodroot, 101
Blue cohosh, 129
Bog *Conocybe*, 152–153
Bog laurel, 87–88
Boric acid, 11
Box elder, 12
Box thorn, 40
Brassicaceae, 51
Brazilian pepper, 24–25
Broccoli, 51
Brussels sprouts, 52
Buckeye, 76, 77
Buckthorn, 117–118
Bulbous buttercup, 109
Burrow weed, 128
Burrow's solution, 12
Bush bean, 56
Buttercup, 13, 27, 108–110, 118
Butterfly weed, 108

Cabbage, 51
Caffeine, 125
Calcium oxalate, 71
California fern, 101
Calla lilies, 68
Candelabra cactus, 96
Candlenut tree, 83
Cannabidolic acid, 33–34
Cannabinaceae, 33
Cannabinoids, 33
Cannabinol, 33
Cape gooseberry, 67
Caper spurge, 97
Capulincillo, 115
Caprifoliaceae, 80
Caraway, 102
Cardol, 19, 23
Cardiac glycosides, 46, 48, 84, 86. See also Glycosides
Carolina jessamine, 112
Carrot, 101–103
Cascara sagrada, 118
Cashew nut tree, 10, 22–23
Cassava, 93
Castor bean, 4, 93
Castor bean plant, 52–54, pl. 9
Cauliflower, 51
Celandine poppy, 99–100
Celastraceae, 114
Celery, 101
Challice vine, 68
Chanterelle, false, 158, pl. 33
Cherry tree, 90–91
Chile pepper, 62

China tree, 78
Chinaberry tree, 78
Chinese lanterns, 67
Christmas rose, 111
Chrysanthemums, 12
Cicutoxin, 104
Climbing lily, 46–47, 49–50
Climbing nightshade, 53–54, pl. 14
Climbing woodbine, 81
Clover, 56
Coban tape, 12
Colchicine, 50
Common buckthorn, 117
Common milkweed, 108
Coniine, 104
Convolvulaceae, 36
Coontie, 77
Coprine, 154–155
Coral berries, 80–82
Coral plant, 79
Coralberry, 110
Corn-lily, 121
Coumarin, 89
Cow-parsnip, 102
Coyotillo, 115–116
Crabs-eye, 57
Cranberry bush, 81
Crape jasmine, 83
Creeping buttercup, 109
Crocus, 46–47, 50–51
Crowfoot, 108
Crown-of-Thorns, 94–95, pl. 21
Cruciferae, 51
Cucumber plants, 13
Curcas bean, 79
Cyads, 77
Cyanide, 6, 56, 60, 61, 90–91, 93
Cyanide Antidote Package, 6
Cyanogenic glycosides, 93
Cyclopeptide, 139
Cytisine, 92

D-Lysergic acid, 37
Daffodil, 47, 49, pl. 7
Daisies, 12
Daphin, 89
Daphne, 88–89
Deadly nightshade, 64–65
Death camas, 46, 106–107
Death's angel, 137–138
Death's cap, 137–138
Delphinine, 120
Dermatitis, 9–27
Destroying angel, 136–137, pl. 27, 28
Devil's apple, 66
Devil's trumpet, 39–40
Devil's walking stick, 74
Diazepam, 104
Digitalis, 48, 54–56, 111
Digitoxins, 55, 84

173

Dogbane family, 83–86, 108
Dogtooth violet, 46
Doll's eyes, 110
Dopamine, 31
Dumb cane, 4, 68, 71
Dumbcain, 68

Eastern fly agaric, 144, pl. 29
Eastern jack o'lantern, 158
Eastern poison oak, 9
Eggplant, 62
Elderberry, 80–82
Elephant's ears, 68, pl. 11
Encephalins, 42
Endorphins, 42
English holly tree, 124–125
English ivy, 73–74
Enzymes, proteolytic, 70
Ergoline, 37
Esters, 66, 122
Euphorbiaceae, 20, 94, 108
European fly honeysuckle, 81
European horse chestnut, 76
European larkspur, 120
European privet, 91

Fabaceae, 56
False Chanterelle, 158, pl. 32
False hellebore, 46, 121–122
False morel, 135, 141–144, 161–162, pl. 30
Fava beans, 56
Federal Bureau of Narcotics, 33
Fern palm, 77
Fiber-head, 149
Florida arrowroot, 77
Florida holly, 24–25
Fly agaric, 144–146
Fly poison, 46
Fool's parsely, 101–104
Foxglove, 54–56
Fruit tree family, 90–91
Furocoumarins, 27

Galitoxin, 108
Gallita bush, 115
Gelsemine, 113
Gelseminine, 113
Gelsemoidine, 113
Gentle cort, 165
Ginseng family, 74
Girdled *Paneolus*, 152–153
Glucoalkoloids, 66
Glucosides, 52, 76, 91, 108
Glycoalkaloid, 63, 122
Glycosides, 60, 82, 87, 89, 93, 111, 115, 118, 125, 129
Golden chain, 56, 92
Golden club, 68
Golden dewdrop, 124
Gourds, 13
Ground Cherry, 67
Ground hemlock, 74
Guelder rose, 81
Gyromitrin, 141–142

Hallucinogen(s)
  history of human use of, 29–30
  intoxication, description of, 28
  mushrooms, 29, 135, 150–154
  plants, 28–44
Hearts-a-bursting, 114
Hederagenin, 74
Helleborein, 111
Hemlock tree, 101
Henbane, 44
Histamines, 26, 70
Hobblebush, 81
Holly, 124–125
Honeysuckle(s), 80–82
Horse chestnut, 5, 76–77, pl. 16
Horse nettle, 65–66
Horseradish, 52
Hyacinth, 46
Hyacinth bean, 56
Hyoscine, 38
Hyoscyamine, 41, 43

Ibotenic acid, 146, 147
Indian cucumber root, 46
Indian poke, 121
Indian tobacco, 111–112
Indole alkaloids, 20, 37
Inky caps, 154–155
Irish potato, 63
Ivy, 73–74

Jack O'Lantern, 158–159
Jack-in-the-Pulpit, 68, 70, pl. 10
Jamestown weed, 37–39
Japanese lacquer tree, 9–10
Japanese pieris, 86
Jatrophin, 79
Jequerity, 57
Jerusalem cherry, 4, 66
Jessamine, 67
Jimmy weed, 128
Jimsonweed, 29, 37–39, pl. 4

Kale, 51
Kelix, 12
Kentucky coffee tree, 56
Kidney beans, 60–61
Kinins, 70, 71

Labrador tea, 86
Laetrile, 91
Lambkill, 87–88
Lantadene, 124
Lantana, 123–124
Larkspur, 118–120
Leguminosae, 56
Lidocaine, 7
Lily family, 46–51
Lily-of-the-Valley, 46–48, pl. 6
Little brown mushrooms (LBM), 135, 139–140
Loasa family, 26

Love bean, 57
Lucky nut, 85–86
Lupines, 56, 58–59, pl. 13
Lycorin, 49
Lysergic acid diethylamide (LSD), 36, 37, 150

Magnesium sulfate, 6
Malabar tree, 94
Manchineel tree, 20–22
Mandrake, 128–129
Mango fruit, 10
Mango tree, 18–19
Manioc, 93
Maple, 12
Marijuana, 29, 32–35, pl. 2
Marsh marigold, 110
Matrimony vine, 40–41
May apple, 128–129, pl. 26
Meadow saffron, 46, 50–51
Mescal bean, 56
Mescaline, 31
Mezereinic acid anhydrase, 89
Mezereon, 88, pl. 20
Milkweed, pl. 24
Milkweeds, 107–108, pl. 24
Mistletoe, 120–121
Monkshood, 118–119
Moonseed, 129–130
Morels, false, 141–144, 161–162, pl. 30
Morning glory, 29, 35–37, pl. 3
Morphine, 42
Mountain laurel, 87–88, pl. 18
Muscarine, 135
Muscimol, 135, 146
Mushrooms
  classification by toxin, 135
  deadly, 136–144
  general precautions, 134–135
  hallucinogenic, 150–155
  parasympathetic stimulators, 144–147
  that are toxic in combination with alcohol, 154–155
  that cause gastroenteritis, 155–167
  that cause sweating, 148–150
Mustard greens, 51
Mustard family, 51–52
Mustard plants, 13
Mycoatropine, 146

Nalaxone, 42
Nanny bush, 81
Nannyberry, 81
Narcipoetin, 49
Narcissum, 49
Narcissus, 4
National Clearinghouse for Poison Control Centers, 132
Nerine lily, 49
Nerioside, 84
Nettles, stinging, 25, pl. 1
Nicotine, 92, 104, 112
Nightshade family, 37, 40–41, 43, 44, 62–67

Norepinephrine, 31
Northern bearded milky, 157
Nutmeg tree, 29

Octapeptides, 138
Oleander, 4, 83–85, pl. 17
Oleandroside, 84
Oleoresin, 16
Oleoresins, 10
Ololiuqui, 35
Opium poppy, 41–42, 72, pl. 5
Oxalic acid, 72
Oxygen therapy, 6

Panther *Amanita*, 146–147
Parsley, 101, 102
Parsnip, 101, 102
Pasqueflower, 110
Peach tree, 90–91
Peanuts, 56
Pencil tree cactus, 94–95
Pepino, 62
Peregrina, 79
Petrolatum, 11
Peyote cactus, 29, 30–32
Phenols, 34
Phenothiazine, 39
Phenytoin, 7
Photosensitizers, 27
Phylloerythrin, 124
Physic nut, 79–80
Physostigmine, 20, 39, 113
Phytolaccine, 105
Phytotoxins, 53, 56, 58, 60, 79
Picrotoxin, 107
Pie-plant, 71
Pigeonberry, 104
Plum tree, 90–91
Podophyllin, 129
Podophylotoxin, 129
Poinciana, 56
Poinsettia, 95, 97
Poison elder, 15
Poison guava, 20
Poison hemlock, 101–104, pl. 22
Poison ivy group, 9–12
Poison ivy, 9, 13–15
  controlling, 16–17
  treatment, 11–12
Poison oak, 9, 14
Poison pie, 159
Poison sumac, 15–17
Poisonwood, 17–18
Pokeberry, 104
Pokeweed, 104–105, pl. 23
Pole beans, 56
Pollen-caused allergic contact dermatitis, 12
Polycyclic diterpenoid alkaloids, 120
Poplars, 12
Potassium permanganate, 11
Potato, 63
Prayer bean, 57
Precatory bean, 57
Prickly poppy, 101

Privet, 91–92
Propranolol, 7
Protoanemonin, 109–110
Psilocin, 150
Psilocybin, 150
Psoralens, 27
Purge nut, 79
Pyridine alkaloids, 112

Queen Ann's lace, 27, 102

Radish, 13, 51
Ranunculaceae, 109
Rattlebox, 56
Rayless goldenrod, 128
Red spurge, 94
Rhododendron family, 86–87, pl. 19
Rhubarb, 71–73
Ricin, 53–54
Rock poppy, 99–100
Rosary pea, 4, 56, 57–58, pl. 12
Rutabaga, 52

St. John's wort, 27
Sandbox tree, 79
Saponins, 24, 48, 66, 74, 79, 82, 105, 118
Sarsaparilla, 74
Scarlet runner bean, 61
Scopolamine, 38
Serotonin, 26
Sesbane, 56
Sheep laurel, 87–88
Shock, anaphylactic, 19
Sierra laurel, 86
Skunk cabbage, 68, 71, 122
Small-flowered buttercup, 109
Snow-on-the-mountain, 96
Snowberries, 80, 81
Snowdrop, 49
Solanaceae, 37, 44
Solanine, 63, 66
Solomon's seal, 46
Spikenard, 74
Spindle tree, 114
Spotted cowbane, 101
Spurge, 5, 13, 20, 108
Spurge cactus, 95
Spurge family, 94–97
Spurge laurel, 88
Spurge nettle, 26
Star of Bethlehem, 46
Steroid alkaloids, 107
Stinging nettles, 25–26, pl. 1
Strawberry bush, 114–115
Strawberry tomato, 67
Strychnine, 112
Swamp sumac, 15
Sweet pea, 56
Syrup of Ipecac, 5

Tall field buttercup, 109
Tanacetin, 126
Tannins, 82
Tansy, 126
Tapioca, 93

Taxine, 75
Tetrahydrocannabinol (THC), 34
Texas umbrella tree, 78
Theobromine, 125
Thevetin, 86
Thorn apple, 37–39, pl. 4
Toggle-switch intoxication, 146, 147
Tomatillo, 67
Tomato, 62
Toxalbumins, 79
Tremetol, 127
Triterpenoid saponins, 74
Tropane, 38
Tropane alkaloid, 41
Trumpet flower, 68
Tullidora, 115
Tung nut, 82–83
Tung-oil tree, 82–83
Turnips, 51
Tyramine, 121

Umbelliferae, 102
Urishol, 9, 10, 14, 15, 16, 18, 19

Veratramine, 122
Viburnums, 80
Virgin's bower, 111
Virginia creeper, 74
Vomiting, how to induce, 5–6

Water hemlock, 101–104
Watercress, 51
Wayfaring-tree, 81
Western poison oak, 9
White snakeroot, 126–128
Wild carrot, 27, 102
Wild lily-of-the-valley, 46
Wild parsnip, 27
Wild raisin, 81
Wild tomato, 65
Wind flower, 110
Withe rod, 81
Woody nightshade, 63–64

Yaupon tree, 125
Yellow jessamine, 112–113
Yellow nightshade, 83
Yellow oleander, 83, 85–86
Yerba mate, 125
Yew, 74–75, pl. 15
Yuca, 93

Zinc oxide, 11
Zygacine, 107

Plants and Mushrooms by Scientific Name
(single name indicates genus)

Abrus precatorius, 56, 57–58
Aconitum columbianum, 119
Aconitum nepellus, 119
Aconitum reclinatum, 119

175

Aconitum uncinatum, 119
Actaea, 110
Aesculus (sp.), 76
Aesculus hippocastanum, 76
Aethusa cynapium, 101
Aleurites fordii, 82–83
Aleurites moluccana, 83
Alocasia, 68
Amanita bisporigera, 137
Amanita chlorinosoma, 166
Amanita citrina, 166
Amanita cothurnata, 147
Amanita flavoconia, 165
Amanita flavonia, 165
Amanita flavorubescens, 165
Amanita gemmata, 166
Amanita muscaria, 135, 144–145, 149
Amanita ocreata, 137
Amanita pantheria, 135, 146–147, 149
Amanita phalloides, 137–139
Amanita verna, 137
Amanita virosa, 136–137
Amanitas, 135, 136–139, 144–147, 165–167
Amaryllis, 47, 49
Amianthium muscaetoxicum, 46
Anacardium occidentale, 10, 22
Andromeda, 86
Anemone, 110
Anthurium, 68
Apocynum, 83
Aralia spinosa, 74
Argemone mexicana, 101
Arisaema triphyllum, 68, 70
Asclepias syriaca, 108
Asclepias tuberosa, 108
Asparagus officinalis, 51
Atropa belladonna, 43

Boletes, 162
Boletus calupus, 163
Boletus eastwoodiae, 162, 163
Boletus luridus, 163
Boletus miniato-divaceus, 163
Boletus satanas, 162
Boletus subvelotipes, 163
Brugmansia, 40

Caladium, 68
Caltha, 110
Cannabis sativa, 29, 32–35
Caulophyllum thalictroides, 129
Celastrus orbiculatus, 114
Celastrus scandens, 114
Cestrum, 67, 112
Chelidonium majus, 99
Cicuta maculata, 101
Clematis, 111
Clitocybe, 135, 148–149
Clitocybe dealbata, 148
Clitocybe dilatata, 148
Clitocybe illudens, 158
Cnidoscolus stimulosus, 26
Colchicum autumnale, 46–47, 50–51

Colocasia, 68
Conium maculatum, 101
Concybe, 29, 135
Conocybe filaris, 141
Concybe smithii, 152–153
Convallaria majalis, 46–48
Convallaria montana, 48
Convolvulus arvensis, 36
Coprinus atromentarius, 154–155
Coprinus comatus, 154
Coprinus micaceus, 154
Coprinus, 135
Cortinarius gentilis, 165
Crinum, 49
Crotalaria, 56
Cycas, 77

Daphne mezereum, 88
Datura, 29
Datura arborea, 40
Datura candida, 40
Datura metel, 40
Datura meteloides, 40
Datura sanguinea, 40
Datura stramonium, 37–38
Datura suaveolens, 39
Daubentonia, 56
Daucaus carota, 27
Delphinium ajacis, 119
Delphinium bicolor, 119
Delphinium cheilanthum, 119
Delphinium grandiflorum, 119
Dieffenbachia, 12, 68–71
Digitalis purpurea, 55
Dioon, 77
Dolichos lablab, 56
Duranta repens, 124

Entoloma, 160
Entoloma sinuatum, 160
Entotoma strictius, 160–161
Entoloma vernum, 160–161
Ervatamia coromaria, 83
Erythronium, 46
Euonymus, 114
Eupatorium rugosum, 126–127
Euphorbia cotinifolia, 94
Euphorbia frankiana, 95
Euphorbia lactea, 96
Euphorbia lathyris, 97
Euphorbia marginata, 96
Euphorbia milii, 94–95
Euphorbia pulcherrima, 95, 97
Euphorbia tirucalli, 94–95

Galanthus nivalis, 49
Galerina, 135, 139–140
Galerina autumnalis, 140, pl. 31
Galerina marginata, 140
Galerina venenata, 140
Gelsemium, 67
Gelsemium sempervirens, 112–113
Gloriosa (sp.), 46, 49
Glottidium, 56
Gymnocladus dioica, 56

Gymnopilus, 135
Gymnopilus spectabilis, 152–153
Gyromitra esculenta, 141–142
Gyromitra infula, 141–142

Haplopappus heterophyllus, 128
Hebeloma crustuliniforme, 159
Hebelomas, 159–160
Hedera helix, 73
Helleborus niger, 111
Helvella esculenta, 141
Helvella infula, 141
Helvella lacunosa, 143–144
Hippeastrum, 49
Hippomane mancinella, 20–21
Hura crepitans, 79
Hyoscyamus niger, 44
Hypericum (sp.), 27

Ilex aquifolium, 124–125
Ilex opaca, 124
Ilex paraguayensis, 125
Ilex vomitoria, 125
Inocybe fastigiata, 149
Inocybe, 135, 149–150
Ipomoea tricolor, 29, 35–36

Jatropha, 79–80
Jatropha curcas, 79
Jatropha gossypifolia, 79
Jatropha hastata, 79
Jatropha intergerrima, 79
Jatropha multifida, 79

Kalmia, 87
Kalmia augustifolia, 88
Kalmia latifolia, 88
Kalmia polifolia, 88
Karwinskia humboldtiana, 115–116
Laburnum anagyroides, 56, 92
Lactarius repraesentaneus, 157
Lactarius rufus, 157
Lactarius scrobiculatus, 158
Lantana camara, 123
Lathyrus, 56
Ledum (sp.), 86
Leucothoe (sp.), 86
Ligustrum vulgare, 91–92
Lobelia cardinalis, 112
Lobelia inflata, 111
Lonicera, 81
Lonicera periclymenum, 81
Lonicera xylosteum, 81
Lophophora williamsii, 29, 30
Lupinus perennis, 59
Lycium balmifolium, 40

Macrozamia, 77
Maianthemum canadense, 46
Mangifera indica, 10, 18
Manihot esculenta, 93
Medeola virginiana, 46
Melia azedarach, 78
Menispermum canadense, 129–130

Metopium toxiferum, 17
Monstera deliciosa, 68–69
Myristica fragrans, 29

Narcissus pseudo-narcissus, 47, 49
Neogyromitra gigas, 143
Nerium oleander, 83–84
Nolina, 46

Omphalotus illudens, 158
Omphalotus olicascens, 158
Ornithogalum, 46
Orontium aquaticum, 68

Panaeolus, 135
Panaeolus subalteatus, 152–153
Panax quinquefolius, 74
Papaver somniferum, 41
Parthenocissus quinquefolia, 74
Pastinaca sativa, 27
Paxillus involutus, 164
Peltandra virginica, 68
Phaseolus coccineus, 61
Phaseolus vulgaris, 56, 60
Philodendron, 12, 68
Phoradendron flavescens, 121
Phoradendron serotinum, 120–121
Phoradendron villosum, 120
Physalis, 67
Phytolacca americana, 104, 105
Phytolacca rigida, 105
Pieris japonica, 86
Podophyllum peltatum, 128
Poinciana silliesii, 56
Polysonatum, 46
Polyscias balfouriana, 24
Prunus serotina, 90
·*Psilocybe*, 29, 135, 150–151

Psilocybe baeocystis, 151
Psilocybe pelicula, 151
Psilocybe semilanceata, 151
Psilocybe strictipes, 151

Ranunculus, 27, 108
Ranunculus abortivus, 109
Ranunculus acris, 109
Ranunculus bulbosus, 109
Ranunculus repens, 109
Rhamnus catharticus, 117–118
Rhamnus frangula, 117–118
Rhamnus purshiana, 118
Rheum rhaponticum, 71
Rhus radicans. See *Toxicodendron radicans*
Ricinus communis, 52–53
Robinia pseudoacacia, 56, 59–60
Russula emetica, 156–177

Sambucus, 81
Sanguinaria canadensis, 101
Schinus terebinthifolius, 24–25
Sesbania, 56
Solandra guttata, 68
Solandra maxima, 68
Solanum aculeatissimum, 66
Solanum americanum, 64–65
Solanum carolinense, 65
Solanum dulcamara, 63–64
Solanum nigrum, 64–65
Solanum pseudo-capsicum, 66
Solanum sodomeum, 66
Solanum tuberosum, 63
Sophora secundiflora, 56
Spathiphyllum, 68
Stropharia, 29
Symphoricarpos, 81
Symphoricarpus albus, 81

Symalocarpus foetidus, 68, 71

Tanacetum vulgare, 126
Taxus baccata, 75
Taxus brevifolia, 75
Taxus canadensis, 75
Taxus floridana, 75
Thevetia peruviana, 83, 85
Toxicodendron diversilobum, 9, 15
Toxicodendron radicans, 9
Toxicodendron toxicarium, 9, 15
Toxicodendron verniciferum, 9
Toxicodendron vernix, 15
Tsuga, 101
Turbina corymbosa, 35

Urechites, 83
Urtica dioica, 25–26

Veratrum, 121
Veratrum californicum, 122
Veratrum parviflorum, 122
Veratrum viride, 46, 122
Verpa bohemica, 161–162
Viburnum, 81
Viburnum lantana, 81
Viburnum opulus, 81
Vicia faba, 56

Wisteria, 56, 61–62

Xanthosoma, 68

Zamia, 77
Zanteschia, 68
Zephyranthes atamasco, 49
Zigadenus, 46
Zigadenus chloranthus, 106

ABOUT THE AUTHORS

Charles Kingsley Levy, Ph.D., has been a park ranger and naturalist at Shenandoah National Park and National Capital Park, as well as a Fulbright professor and Reader in Zoology at the University of East Africa, Kenya. Levy has written ten bestselling textbooks in biology and over 50 research publications in his field. Currently professor of biology at Boston University, he continues independent research in the field of environmental physiology. Among his interests are wildlife photography, African art, bird watching, and shell collecting. He resides in Brookline, Massachusetts.

Richard B. Primack, Ph.D., has unusually wide botanical experience from his many travels—including the mountains of New Zealand and Chile, the coastal swamps of Australia, the rain forests of Costa Rica, and the tropical forests of Borneo, where he met his wife, Margaret. He has published over 40 scientific papers in plant ecology and is currently an Assistant Professor at Boston University. His interests include cultural anthropology, Chinese cooking, and botanical humor. He alternates his residence between Boston and Borneo.

L. Laszlo Meszoly teaches scientific illustration at the Harvard Museum of Comparative Zoology. His work has appeared on the covers of *Science* and *Nature*, as well as in numerous other scientific journals and books. He contributed all the mushroom drawings to this book.

Margaret Huong Primack is an accomplished batik artist who brought a delicate and new touch to the plant line drawings in this book.